MICROBIAL AND MOLECULAR GENETICS

THE GENETIC CODE

First base	Second base: U		Second base: C		Second base: A		Second base: G		Third base
U	UUU	Phenylalanine	UCU	Serine	UAU	Tyrosine	UGU	Cysteine	U
	UUC	Phenylalanine	UCC	Serine	UAC	Tyrosine	UGC	Cysteine	C
	UUA	Leucine	UCA	Serine	UAA	'Ochre'	UGA*		A
	UUG	Leucine	UCG	Serine	UAG	'Amber'	UGG	Tryptophan	G
C	CUU	Leucine	CCU	Proline	CAU	Histidine	CGU	Arginine	U
	CUC	Leucine	CCC	Proline	CAC	Histidine	CGC	Arginine	C
	CUA	Leucine	CCA	Proline	CAA	Glutamine	CGA	Arginine	A
	CUG	Leucine	CCG	Proline	CAG	Glutamine	CGG	Arginine	G
A	AUU	Isoleucine	ACU	Threonine	AAU	Asparagine	AGU	Serine	U
	AUC	Isoleucine	ACC	Threonine	AAC	Asparagine	AGC	Serine	C
	AUA	Isoleucine	ACA	Threonine	AAA	Lysine	AGA	Arginine	A
	AUG	Methionine	ACG	Threonine	AAG	Lysine	AGG	Arginine	G
G	GUU	Valine	GCU	Alanine	GAU	Aspartic acid	GGU	Glycine	U
	GUC	Valine	GCC	Alanine	GAC	Aspartic acid	GGC	Glycine	C
	GUA	Valine	GCA	Alanine	GAA	Glutamic acid	GGA	Glycine	A
	GUG	Valine	GCG	Alanine	GAG	Glutamic acid	GGG	Glycine	G

Notes

1. *UGA, like 'ochre' and 'amber', brings about chain termination.
2. The trinucleotide sequences are written left to right from the 5′ to the 3′ direction. The anticodon sequences in tRNA, written in the same way, pair with the codons in reverse order; i.e. the anticodon bases 1, 2 and 3 pair with the codon bases 3, 2 and 1 respectively.

From Fincham and Day, *Fungal Genetics*, Blackwell Scientific Publications Ltd., 1971.

MICROBIAL AND MOLECULAR GENETICS

2nd edition

J. R. S. FINCHAM, Sc.D., F.R.S.

Professor of Genetics,
University of Edinburgh
formerly University of Leeds

HODDER AND STOUGHTON
LONDON SYDNEY AUCKLAND TORONTO

CRANE, RUSSAK & CO. INC., NEW YORK

BIOLOGICAL SCIENCE TEXTS

General Editor
Don R. Arthur, M.Sc., Ph.D., D.Sc., F.I.Biol.
Professor of Zoology, King's College
University of London

BIOLOGY, FOOD AND PEOPLE
Robert Barrass, B.Sc., Ph.D., F.I.Biol.

HISTORY OF BRITISH VEGETATION
Winifred Pennington, B.Sc., Ph.D.

FOOD SCIENCE
Brian A. Fox, B.Sc., F.R.I.C., F.I.F.S.T.
& Allan G. Cameron, B.Sc., F.R.I.C., F.I.F.S.T.

SURVIVAL
Don R. Arthur, M.Sc., Ph.D., D.Sc., F.I.Biol.

CELL RESPIRATION
W.O. James, F.R.S.

First edition 1965
Second edition 1976

Hodder and Stoughton Educational,
a division of Hodder and Stoughton Limited, London

ISBN 0 340 18067 6 Case Bound Edition
ISBN 0 340 18068 4 Paper Bound Edition

Published in the United States by:
Crane, Russak & Company Inc.
347 Madison Avenue New York NY 10017

ISBN 0 8448 0768 0 Case Bound Edition
ISBN 0 8448 0769 9 Paper Bound Edition
Library of Congress Catalog Card Number 75 21729

Computer Typesetting by Print Origination, Orrell Mount
Hawthorne Road, Bootle, Merseyside L20 6NS
Printed in Great Britain by
J. W. Arrowsmith Limited
Winterstoke Road, Bristol BS3 2NT

Preface

The very extensive advances which have been made in molecular genetics in the ten years which have elapsed since the first edition of this book was published have necessitated considerable rewriting throughout and some changes of emphasis. In particular, the chapter on *Episomes* has been largely rewritten and retitled *Plasmids,* while a new chapter has been added on *Physical Mapping and Manipulation of the Genome* to take account of work which has added a whole new dimension to genetics. I still hope that the book will not be beyond the ambitious beginner in biology but I have continued to include selected references to original papers which may serve as leads into the literature for more advanced students.

In the Preface to the first edition, written at the John Innes Institute, I acknowledged my debt to my then colleague, Dr. Robin Holliday, for his critical reading of the manuscript. This time I must thank my colleagues at Leeds University, Drs. Simon Baumberg and Sheena Dennison, for valuable advice on specific points and for providing an atmosphere of continuing discussion which has helped me, I hope, to keep reasonably up-to-date.

University of Leeds

J.R.S. FINCHAM

Contents

Preface v

1 **Introduction** 1
Basic problems of genetics. Genetic systems in eukaryotes. Special features of micro-organisms (fungi, bacteria, viruses).

2 **The Chemical Nature of the Genetic Material** 10
Nucleic acids and proteins. The self-replication of DNA. DNA replication in a cell-free system. Genetic transformation with DNA. The genetic material of viruses.

3 **Mapping the Genome** 28
Introduction. *Neurospora*—an organism with a regular sexual cycle. Conjugation and recombination in *Escherichia coli.* Transduction—genetic transfer through virus infection (generalised transduction, specialised transduction). Mapping by transformation with DNA. Genetic mapping in bacteriophages. The universality of genetic recombination and its mechanism. A comparison of prokaryote and eukaryote chromosomes.

4 **Mutation** 55
Spontaneous mutation. Selection of auxotrophic mutants. Conditional mutants. Reverse mutation. Mutagens and the mechanisms of mutation. Properties of individual mutational sites.

5 **Gene Action** 71
Enzymes. The genetic unit of function—the cistron or gene. Complementation tests in different organisms. One gene—one enzyme (histidine mutants of Salmonella, tryptophan mutants of *E. coli*). Intra-genic complementation. Biochemical evidence on the mechanism of synthesis of RNA and proteins (transcription of DNA into RNA, ribosomes, messenger RNA, transfer RNA, the process of polypeptide synthesis, the code). Genetic confirmation of the code (recombination within a codon, successive mutations in the same codon, evidence from frameshift mutations, double-frameshift analysis, chain-terminating mutants, normal chain termination, mutations in initiating codons).

6 **Regulation of Gene Action** 92
Controls of quantity and controls of activity—allosteric proteins. Induction and repression. The *lactose* mutants of *Escherichia coli* (the structure-determining genes, constitutive mutants—repressor and operator, the promoter region and initiation of transcription, nucleotide sequence of the operator-promoter region). Other inducible enzyme systems—the *ara* operon. The relation between repressible and inducible systems. The histidine operon of *Salmonella typhimurium* (co-ordinate regulation, the nature of the repressor, the operator). Polarity within operons. The general significance of operons and of close linkage. Regulation of transcription during development.

7 **Plasmids** 111
Independent replication of plasmids. Physical form of plasmid DNA. Capacity for transfer. Integration into the bacterial chromosome. Additional gene functions of plasmids—colicins and resistance. Origin and evolution of plasmids.

8 **Physico-chemical Mapping and Manipulation of the Genome** 121
Hybridisation of nucleic acid molecules. Physical mapping using the electron microscope. Specific enzymic fragmentation of DNA and ordering the fragments. Possibilities of fractionation of the eukaryote genome. Visualisation of specific genes. Gene isolation. Chemical determination of gene structure and gene synthesis. The construction of artificial genetic vectors.

References and Further Reading 139

Index 145

1 Introduction

BASIC PROBLEMS OF GENETICS

Genetics has been described as the study of *variation* and its inheritance. It would be equally true to say that the subject is concerned with the mechanisms underlying the high degree of *constancy* shown by biological species. While striking variations do occur, and form the raw material for genetic experimentation, the very fact that we are able to recognise well-defined species implies that living things are able to transmit the capacity for a particular type of complicated development from generation to generation with extreme accuracy. This seems all the more remarkable when we remember that, in higher plants and animals, the only physical link between successive sexually produced generations is a single cell, the fertilised egg, which itself shows no signs of the various and complex structures which will arise from its development. Clearly, a fertilised egg must contain a vast amount of information which can be transmitted accurately through an indefinite number of cell divisions and guide the organism through all the contingencies of growth and cellular differentiation.

In spite of their small size and apparently simple structures, bacteria and other types of micro-organism must also possess the capacity for storing and transmitting genetic information. As we shall see in Chapter 6, bacteria are extremely versatile organisms, and are capable of making drastic changes in their metabolism to meet alterations in their chemical environment. These metabolic adjustments depend to a large extent on the ability of the bacterial cell to discontinue the production of certain *enzymes* (specific catalytic proteins) and to start the production of others. However, the organism 'remembers' how to form the proteins which it has stopped making, and is able to revert to the original pattern of protein synthesis immediately on being returned to the original environment. Like the cells of higher organisms, the bacterial cell must carry more genetic information than it is actually acting upon at any given time. What the controls are which determine that a cell shall express only a part of its genetic potential in a given situation is perhaps the most important question in modern biology. In relation to higher organisms, this is the problem of cellular differentiation, and of cancer, which can be regarded as a pathological form of differentiation. It is, however, chiefly in bacteria that the mechanisms are now at least partly understood.

This book will be concerned with three fundamental questions, and with the experimental systems used for investigating them:

(i) What is the chemical basis of the storage of genetic information in the cell, or, to put it another way, what is the nature of the genetic material?

(ii) In what kind of code is the genetic information written, and how is it translated into action in the form of specific cellular process?

(iii) If the whole of the genetic information of a cell is not constantly in use, what is it that determines which part of it shall be 'switched on' and which 'switched off'?

If it were not for the genetic study of micro-organisms very little could be said about any of these questions. The work which has been done during the past 25 years on viruses, bacteria, and, to a lesser extent, on filamentous fungi provides some clear answers to these questions, and suggests working hypotheses even where certain information is lacking.

GENETIC SYSTEMS IN EUKARYOTES

Classical genetics is based on experiments with animals, vascular plants, fungi and typical algae. These groups, collectively known as the *eukaryotes,* characteristically possess a precise and regular mechanism—*sexual reproduction*—for combining and reassorting their genetic material. As we shall see in Chapter 3, it is through making controlled matings, and studying the modes of reassortment of genetic factors, that one can study the structure of the genetic material. It seems appropriate here, in order to put the special features of micro-organisms into perspective, to give a brief account of 'orthodox' sexual reproduction.

Eukaryotes mostly have a cellular structure, the main exceptions being certain fungi and algae in which the living material is a continuum of filaments not divided, or only occasionally divided, by cell walls or *septa.* In cellular forms each cell contains a single *nucleus,* which divides when the cell divides. In non-cellular (*coenocytic*) forms there are usually several or many nuclei sharing a common domain, which divide to keep pace with the growth of the coenocyte. In either case the nucleus has basically a common structure and mode of division (*mitosis*). It is separated from the *cytoplasm*—the rest of the contents of the cell or coenocyte—by a nuclear membrane, and is filled by a definite number of thread-like *chromosomes* which, as we shall see, are the carriers of the genetic information. Mitosis involves the longitudinal division of each chromosome with the subsequent separation of each pair of sister chromosomes into different daughter nuclei. Thus each nucleus formed by a mitotic division has the same standard set of chromosomes as the nucleus which gave rise to it.

During the greater part of the life cycle of most higher organisms there are two chromosomes of each kind in each nucleus; the two members of each pair are similar to each other in length and in function, but generally differ from other pairs in both respects. This state of having a double chromosome set is called the *diploid* condition.

In all sexually reproducing organisms the diploid phase is initiated by the process of fusion of pairs of germ cells or *gametes,* with accompanying fusion of nuclei. The gametes are *haploid,* that is to say each has only a single set of chromosomes. Thus the double chromosome set of a diploid organism consists of one set of maternal and one of paternal origin. For germ cells to be formed by the diploid organism the chromosome number must be halved, with the sorting out of complete haploid sets from the diploid complement of chromosomes. The process through which this is accomplished is called *meiosis.*

In animals the haploid phase of the life cycle does not involve any mitotic divisions of haploid nuclei; the products of meiosis develop directly into the gametes. In plants, where the products of meiosis are generally called *spores* (though in fungi the same name is sometimes given to the agents of asexual propagation), the gamete nuclei are formed only after a number of mitotic divisions. In flowering plants this haploid mitosis is restricted to a very few divisions, often three in the embryo sac to give eight haploid nuclei including the egg nucleus, and two in the germinating pollen grain giving the tube nucleus and the two gamete nuclei. In other groups of plants the haploid phase is much more protracted. In ferns, for example, the germinating spore gives by successive mitotic divisions a small haploid green plant, the *prothallus,* on which the gametes are produced. In mosses the entire leafy plant is haploid, and only the capsule formed after fusion of gametes is diploid. Most fungi are at the opposite extreme from flowering plants in being haploid for almost their entire life cycle, meiosis following immediately after nuclear fusion. The yeasts, which are often stably diploid, are the best-known exceptions to the general haploidy of fungi. Fig. 1 shows in generalised form the course of the sexual cycle in a plant.

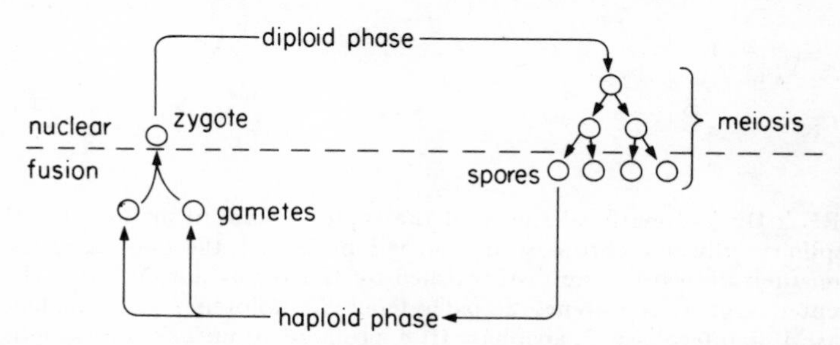

FIGURE 1 Generalised diagram of the life cycle of a plant with regular sexual reproduction.

Meiosis. Although different sexually reproducing organisms may differ greatly in chromosome number, and fungi, for example, have chromosomes which are very minute compared with those of higher plants and animals, the process of meiosis seems to be essentially the same wherever it occurs. It can be most simply described as one division of the chromosomes occurring during *two* divisions of the nucleus, with the result that four haploid nuclei are formed from one original diploid nucleus.

During the first division of meiosis the pairs of similar chromosomes become closely associated point-for-point along their lengths. This capacity for *homologous pairing* is a general characteristic of genetic material and its mechanism is not really understood. It is probable that association through hydrogen bonding of complementary single strands of deoxyribonucleic acid (see p.12) occurs afters the homologous chromosomes are closely associated. In eukaryotes at least, this close association is stabilised by the formation of a predominantly protein structure between the homologous chromosomes; this interesting structure, clearly seen with the electron microscope, is called the *synaptonemal complex.* During the paired phase (*pachytene*) the chromosomes (which have already replicated internally before the onset of meiosis) become visibly double, and the resulting half-chromosomes, or *chromatids,* can (at least in favourable material with large chromosomes) be seen to have 'crossed-over' or exchanged partners at one or

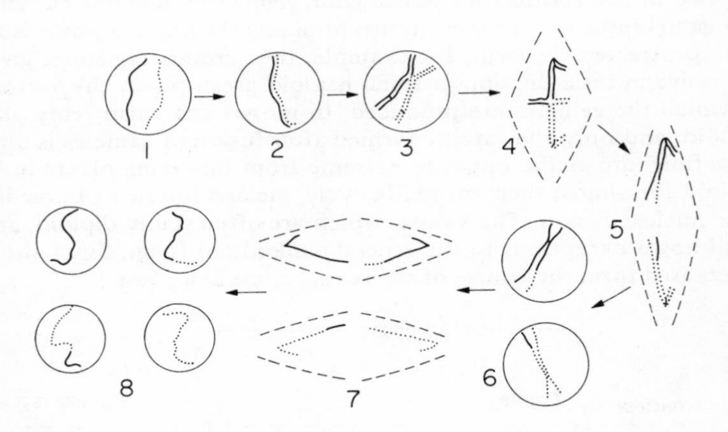

FIGURE 2 Diagrammatic summary of the typical course of meiosis. For the sake of simplicity only one chromosome pair is represented; the two members of the pair and their derivatives are distinguished by full versus dotted lines. The stages represented are: 1, leptotene; 2, pachytene; 3, diplotene; 4, metaphase I; 5, anaphase I; 6, interphase; 7, anaphase II; 8, products of meiosis. In metaphase and anaphase the membrane enclosing the nucleus disappears and is replaced by a system of protein fibres, the spindle, which is involved in chromosome movement and is here indicated in outline (dashed lines). The point at which the two halves of each divided chromosome are held together until anaphase II, and which leads the way to the spindle poles at anaphase, is called the centromere. Note the exchange of chromosome segments which occurs as a result of the chiasmata visible at diplotene.

more points along the length of each chromosome pair. These cross-over points are called *chiasmata* (chiasma in the singular). Soon after this stage (*diplotene*) the chromosome pairs, joined by their chiasmata, all come to lie on the equatorial plane of a *spindle*-like structure, composed of protein fibres, with the two chromosomes of each pair tending to be pulled in opposite directions towards the two apices (poles) of the spindle. This stage, which is called *metaphase I*, is shortly succeeded by *anaphase I* in which the two divided chromosomes of each pair are separated and pulled apart towards the spindle poles.

The second division of meiosis, which follows close upon the first, does not involve any further chromosome division and simply separates, by a second metaphase and anaphase (metaphase II and anaphase II), the two halves of each divided chromosome. Fig. 2 gives the whole process in outline, and shows how, as a result of crossing-over of chromatids prior to diplotene, the haploid products may contain chromosomes combining segments descended from different parents.

SPECIAL FEATURES OF MICRO-ORGANISMS

Fungi

The micro-organisms that have been used in the kind of genetics with which this book is concerned are a very heterogeneous group. However, most of them (with the major exception of certain yeasts, including common bakers' or brewers' yeast, *Saccharomyces cerevisiae*) are normally haploid for essentially the whole of their life histories. Haploidy is a great advantage for experimental genetic studies, since any genetic change (*mutation*) in a haploid has a good chance of showing in the observable characteristics (the *phenotype*) of the organism. In a diploid the effects of a great many mutations are masked because of the unchanged function of the duplicate chromosome—one non-mutant chromosome out of two is very commonly sufficient to maintain a normal phenotype.

Fungi, of which *Neurospora crassa* and *Saccharomyces cerevisiae* are the species which have been the most used by geneticists, are much more akin to higher organisms than to bacteria in their systems of sexual reproduction. The life history of *N.crassa* is shown in diagrammatic form in Fig. 3. It is not necessary to describe in detail all the developmental processes and structures involved. It should, however, be noted that a perfectly orthodox meiosis occurs in the developing *ascus*, which is the only diploid cell in the entire life cycle. Following meiosis the four resulting nuclei each undergo a further mitotic division, so that the ascus comes to contain eight haploid nuclei around which eight *ascospores* are formed. After release from the ascus the ascospores may be cultured individually, but a *N.crassa* strain derived from a single spore is always sexually sterile. The flask-shaped *perithecia*, which contain the asci, are only formed in a mixed culture containing both *mating types* of the fungus. Single ascospores are always either of one mating type or of the other, but the two nuclei, whose fusion initiates the development of each ascus, must be of different mating type. The two mating types are designated *A* and *a*.

One important respect in which *N.crassa*, in common with many other fungi, differs from higher organisms is that it does not, strictly speaking, have a cellular organization. The organism in its vegetative phase consists of branched filaments which grow at their tips and constantly form new branches. These filaments, called *hyphae*, are subdivided by cross walls (*septa*), but the compartments between the septa each contain many nuclei, often of the order of a hundred. Furthermore, each septum has a central pore through which the hyphal contents, including nuclei, can pass. The whole branching hyphal system (the *mycelium*) is a continuum, and living material, including nuclei, tends to flow towards the growing hyphal tips. To increase the fluidity of the situation still further, hyphae of different strains, provided they are not too distantly related, can readily undergo fusions with consequent intermingling of their contents.

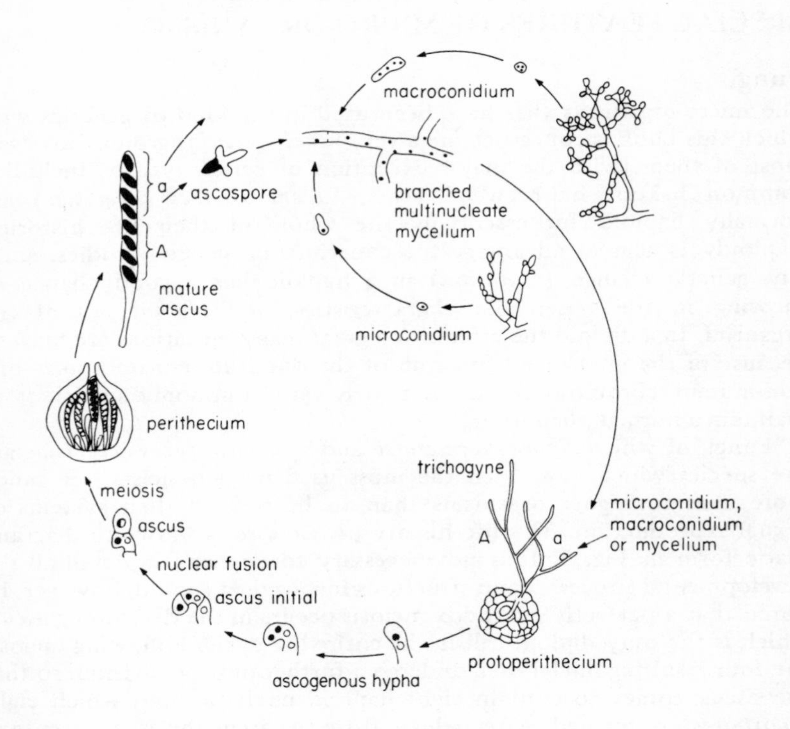

FIGURE 3 Diagrammatic summary of the life cycle of *Neurospora crassa*. From Fincham and Day, ref. p. 140.

The sexual life cycle of *Saccharomyces cerevisiae* is similar to that of *N.crassa* in the alternation of haploid and diploid phases, punctuated by nuclear fusion and meiosis, but is much simpler morphologically. Free cells with single nuclei propagate by budding. The diploid phase is initiated by fusion of free haploid cells of opposite mating type, and meiosis leads to the formation of four haploid cells (*ascospores*) within the persistent cell wall of the diploid mother cell (*ascus*). On their release from the ascus the ascospores bud, if separated from each other, to give stable haploid cultures, but if allowed to remain together cells of different mating type readily fuse to reconstitute the diploid phase. Unlike the situation in Neurospora and most other fungi, diploid nuclei in yeast[1] can divide by mitosis and a diploid culture can multiply by budding indefinitely. Meiosis and formation of ascospores can, however, be induced at will by subjecting a diploid culture to starvation conditions. Fig. 4 summarises the life history of *Saccharomyces cerevisiae.*

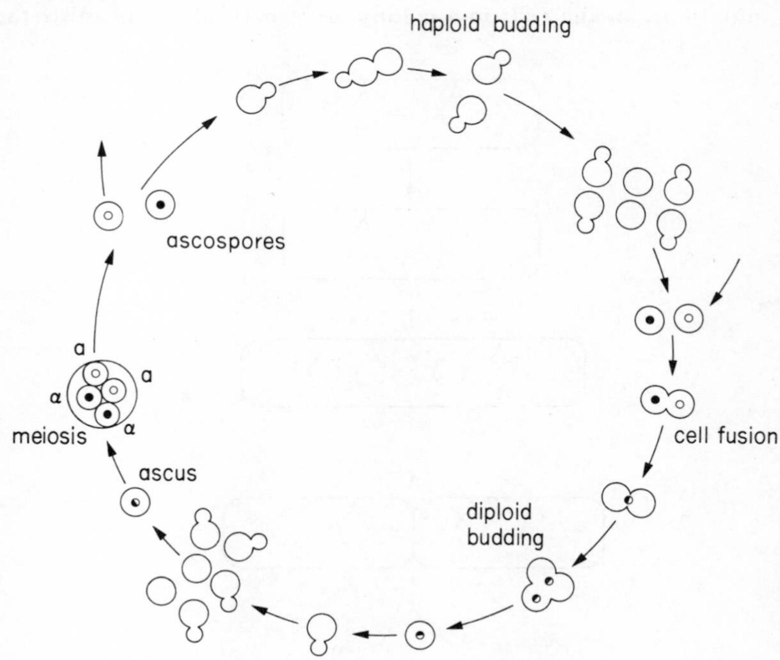

FIGURE 4 Life cycle of *Saccharomyces cerevisiae* (adapted from Fincham and Day, ref. p. 140). Nuclei of the different mating types *a* and α are indicated in relevant cells by open and closed circles; diploid nuclei hybrid for mating type by half-filled circles.

[1] 'Yeast' in this book means *Saccharomyces cerevisiae*, though there are several other genera of yeast-like fungi.

Bacteria

Although Neurospora, with its hyphae of about 5μm (5×10^{-6} m) in width and of yeast, with its cells of about 5 μm in diameter, are often referred to as micro-organisms, bacteria are very much smaller. The rod-shaped organisms of the group Enterobacteriaceae, of which the ubiquitous and usually harmless intestinal inhabitant *Escherichia coli*, and the mouse typhoid organism *Salmonella typhimurium*, have been the most extensively studied by geneticists, are only about 1-2 μm in length, and less than 1 μm in diameter.

The study of bacteria (and the same applies to yeasts) is made a good deal easier by the technique of *plating*, that is, the spreading of a small volume of a dilute suspension of cells over the surface of nutrient medium gelled with agar and contained in a Petri dish. If the medium is a suitable one, each cell divides on the plate to form a small circular colony which can easily be seen with the unaided eye. Each colony contains many thousands of cells, but, in the usual case, and apart from rare mutations, all the cells in a colony are genetically identical to the

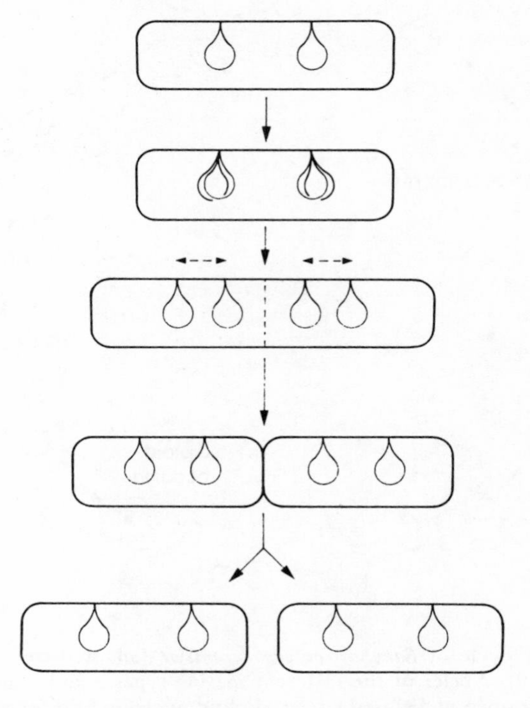

FIGURE 5 The mode of division of a rod-shaped bacterium such as *Escherichia coli* or *Bacillis subtilis*. The DNA is a closed loop (much longer and more elaborately folded than can be shown in a diagram) attached to the cell membrane, the growth of which, according to one interpretation, carries the products of DNA replication apart. See Ryter *et al.* (1968).

single cell which gave rise to them. Thus, although individual bacteria are troublesome to handle, single colonies which represent them many times multiplied are easy to count and test in various ways.

Each rod of *E.coli* or *S.typhimurium* contains two to four 'nuclear bodies', which correspond to the nuclei of larger organisms, though they are very much smaller and different in their organization (see Chapter 3). These bodies are at the extreme limit of the powers of resolution of the light microscope, and it is indirect genetic evidence, rather than direct microscopic observation, which tells us that each consists of a single piece of genetic material analogous to a chromosome (cf. pp. 34-6). The several bodies in one bacterium are all identical, having arisen from a common ancestor by division. Thus a bacterial cell of the *E. coli* type is not equivalent to a uninucleate cell of a higher organism, but rather to a small group of cells with a recent common ancestor. The nuclear bodies, which are really closed loops of tightly packed and elaborately folded DNA (see Chapter 2), divide once during the bacterial division cycle, and are distributed between the two products of each division in an equal and regular way (Fig. 5).

Whereas in sexually reproducing fungi, as well as in higher organisms, there is a completely diploid stage in the life cycle, in bacteria the formation of a completely diploid cell almost never occurs, so far as is known. There are, however, a variety of processes, which will be described in Chapter 3, which result in the formation of partial diploids, an *incomplete* set of genetic material being introduced from one cell into another.

Viruses

Viruses are several orders of magnitude smaller even than bacteria, and the smallest of them have dimensions which are measured in nanometres (i.e. units of 10^{-9} m), and have particles which consist of quite small numbers (e.g. a few hundreds) of large molecules. They are not living organisms in their own right, but are merely devices for inducing living cells to produce more virus of the same kind. Nevertheless they have a genetics of their own, a consideration of which will be deferred until Chapters 3 and 6. Viruses attacking bacteria are called *bacteriophages*.

2 The Chemical Nature of the Genetic Material

NUCLEIC ACIDS AND PROTEINS

If one were looking, without preconceptions, for the repository of genetic information in the living cell, it would be logical to concentrate attention on the *proteins* and *nucleic acids*. Only these components of the cell possess the degree of complexity which one would think necessary for specifying all the details of structure and metabolism which distinguish one biological species from another. Both proteins and nucleic acids consist of long sequences of different kinds of unit, like the sequence of letters in a sentence. In the case of proteins the units are twenty different kinds of *amino acids*, while in nucleic acids they are four different kinds of *nucleotides*. We shall deal with these two classes of macromolecules in turn.

Nucleic acids

There are two quite distinct kinds of nucleic acid–RNA and DNA. In both types the units, the nucleotides, each consist of a nitrogenous base linked to a pentose (5-carbon) sugar phosphate. In *ribonucleic acid* (RNA) the pentose is *D-ribose*, and there are four kinds of base, the two purines *adenine* and *guanine* and the two pyrimidines, *cytosine* and *uracil*. In some kinds of RNA, especially transfer- or t-RNA (see Chapter 5) there are important quantities of minority bases of other kinds, but these need not concern use in the present context. In *deoxyribonucleic acid* (DNA) the pentose is *deoxyribose* rather than ribose, and uracil is replaced by its methylated derivative *thymine*. As in RNA there are other bases occurring in small amounts in some DNA molecules, but the only variation in the general pattern that will be relevant to the subject matter of this book is the occurrence of hydroxymethyl cytosine instead of cytosine in the bacterial viruses (bacteriophages) T2 and T4. The general structure of a polynucleotide chain is shown in Fig. 6, and the structures of the more important nucleic acid bases in Fig. 7. The presence of the free hydroxyl group on the ribose of RNA but not on the deoxyribose of DNA causes some important differences in chemical properties, notably the extreme sensitivity of RNA to mild alkaline hydrolysis in contrast to the resistance of DNA to the same treatment. DNA and RNA also differ in being attacked and hydrolysed by different classes of specific enzymes, called DNAase and RNAase respectively.

In eukaryotes, where a clear distinction can be made between nucleus and cytoplasm, it is known that most of the DNA is in the nucleus, with minor amounts in other cell organelles such as *mitochondria* and *plastids.* RNA, on the other hand, occurs in major amounts in the cytoplasm as well as in the nucleus. While the nucleus-cytoplasm distinction is not strictly applicable to bacteria, so-called *nuclear bodies* (Fig. 5) do occur in bacterial cells and contain at least the great bulk of the DNA (though they are not enclosed in nuclear membranes) while the RNA is more generally distributed.

Several kinds of RNA with different functions are now recognised, and these range in molecular weight from about 20 000 (about 80 nucleotides) for t-RNA to a million or more. The complete nucleotide sequences of numerous different kinds of t-RNA have now been worked out. DNA molecules, on the other hand, are nearly always very much larger—exactly how large is often difficult to determine since the enormous and rather fragile molecules usually get broken up during the process of isolation from the cell. It seems clear, for example, that all the DNA in one bacterial nuclear body is in one piece of a molecular weight of approximately 10^{10}, but most preparations of purified bacterial DNA contain fragments of no more than 10^8 daltons[1]. In yeast, which has the smallest eukaryotic chromosomes known, the DNA in each chromatid consists of one continuous molecule; whether the same is true in higher eukaryotes is less clear, but seems likely.

A key observation for the elucidation of the structure of DNA was that of Chargaff, who showed that, while DNA isolated from different organisms had different, and characteristic, base ratios, the adenine content was always more or less equal, on a molar basis, to the thymine content, and the guanine content was equal to the cytosine content. Subsequent work based on X-ray diffraction led Watson and Crick to propose the now generally accepted double helical structure. In this structure two poly-nucleotide strands are coiled together, each being wound around the central axis of the molecule in a helix. The two strands are also wound around each other, so that they cannot be separated without uncoiling. As will be seen in Fig. 6 a polynucleotide strand is *polarised,* the number 5 carbon atoms of the pentose residues being directed towards one end of the strand, and the number 3 carbon atoms towards the other. In the Watson-Crick model the polarities of the two strands run in opposite directions. The bases, which are attached as side chains to the main pentose-phosphate strands, are directed inwards to the central axis of the double helix, and the two strands are held together by the attachment, through weak bonds involving sharing of hydrogen atoms (*hydrogen bonds*), of each base on one chain with the base at the same level on the partner chain. The most significant feature of the structure is that only two kinds of base pair are geometrically possible, adenine-thymine and guanine-cytosine. Fig. 7 shows these two

[1] The dalton, the mass of the hydrogen atom, is the unit of molecular weight.

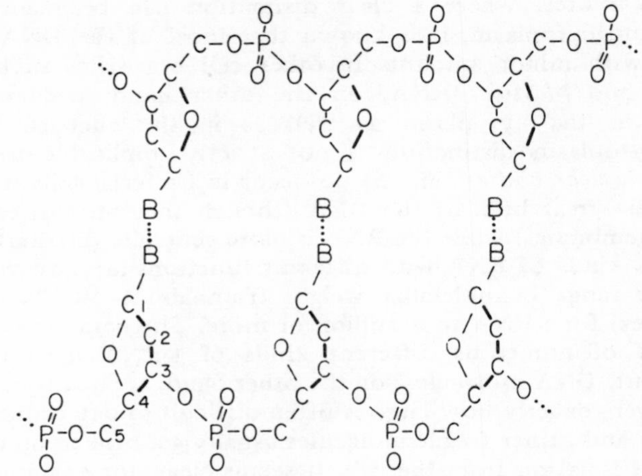

FIGURE 6 Short sections of two DNA chains paired as in a double-stranded molecule. Bases are indicated by B, and hydrogen bonding between bases is indicated by dotted lines. Note that the two strands run in opposite directions. Hydrogen atoms, which fill the remaining carbon valences, are omitted from the diagram for the sake of simplicity. In RNA the number 2 carbon of the ribose ring (see numbering on the lower left ribose residue) carries a free hydroxyl group.

base pairs, and Fig. 6 will give an idea of the structure of a section of the double-stranded molecule. The important point to note is that, since each base will pair with only one other, the sequence of bases along one strand defines a unique sequence of complementary bases along the other. The significance of this for the biological synthesis of DNA will emerge later in this chapter.

As we shall see in Chapter 5 it is more difficult to give a concise account of RNA because there are several functionally distinct kinds differing from each other in molecular size and structure. Although the RNA bases, like those of DNA, are capable of forming specific

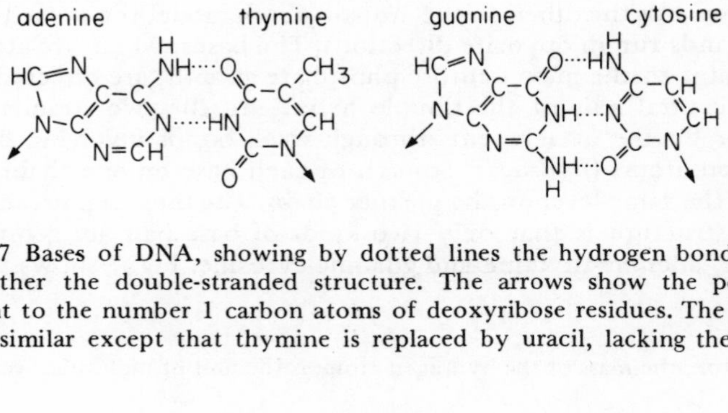

FIGURE 7 Bases of DNA, showing by dotted lines the hydrogen bonds which hold together the double-stranded structure. The arrows show the points of attachment to the number 1 carbon atoms of deoxyribose residues. The bases of RNA are similar except that thymine is replaced by uracil, lacking the methyl group.

hydrogen-bonded pairs, uracil substituting for thymine, RNA in general does not have a regular two-stranded structure, though a single strand can and often does fold back on itself to form a hairpin loop stabilised by base pairing. The three-dimensional structure recently worked out for one kind of *t*-RNA (see pp. 83) shows a complex pattern of such hydrogen-bonded loops.

Proteins

Proteins are the most complex of the large molecules found in living cells, and they account for the largest proportion of the mass of a typical cell. Proteins are built out of *polypeptide chains,* and each such chain is formed by the condensation of (usually) a few hundred L-α-amino acid molecules; the D optical isomers do not occur in typical proteins. There are twenty different amino acids occurring in proteins; a few special ones, such as hydroxyproline which occurs in collagen of muscle, seem always to be formed by secondary modification of one of the primary twenty.

Fig. 8 shows the general structure of a polypeptide chain, and Figs. 8 and 9 the structures of the standard set of 20 amino acids. One of these, proline, is not really an amino acid at all but rather an *imino* acid, with the nitrogen forming part of a five-membered ring instead of providing a free amino group. The sequence of amino acid residues along the polypeptide chain constitutes the *primary structure* of a protein.

Since there seem to be no restrictions on the sequences which are possible it will be seen that, even for a polypeptide chain of the relatively small size of 100 amino acids, there is an exceedingly large number (20^{100}) of possible sequences.

The primary structure, however, is only the beginning of the complexity of a protein molecule. The polypeptide chains can be coiled in various ways, for example in the form first described by Pauling and Corey and known as the *alpha helix.* In the Pauling-Corey structure a regular helical coiling of the polypeptide chain is held in place by hydrogen bonds formed between imino (>NH) and keto (>C=0) groups. Many proteins contain considerable stretches of alpha-helical structure, but equally important may be hydrogen-bonded associations of paired sections of polypeptide chains in either anti-parallel or parallel orientation (the so-called beta or pleated sheet structure). Another complication is that certain polypeptide chains have disulphide (–S--S) bridges formed by oxidation between cysteine residues which may be quite far apart in the primary structure.

The first-order folding of a polypeptide chain, as in an alpha-helix, is called *secondary structure,* and second-order folding (coiling of the coil) is called *tertiary structure.* Added to all this is the fact that a great many proteins, including, probably, nearly all the larger ones of molecular weight of 10^5 or more, consist of *aggregates* of polypeptide chains. Many protein molecules consist of 2, 4, 6 or even more polypeptide chains, frequently identical but sometimes of more than one kind. The bonds which hold these aggregates together are probably mainly of the

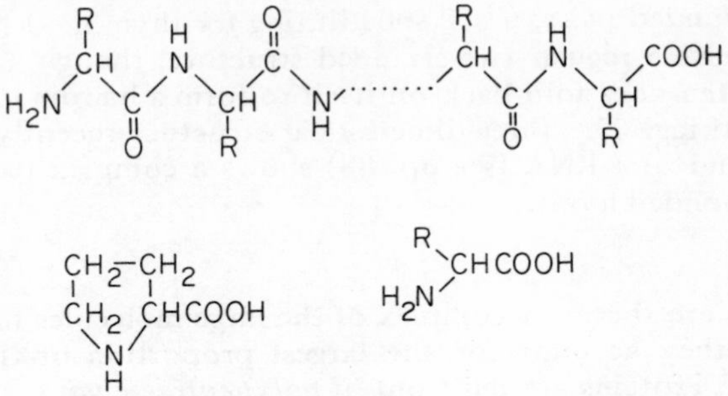

FIGURE 8 General structure of an α-amino acid (lower right) and proline lower left) and of a polypeptide chain (above).

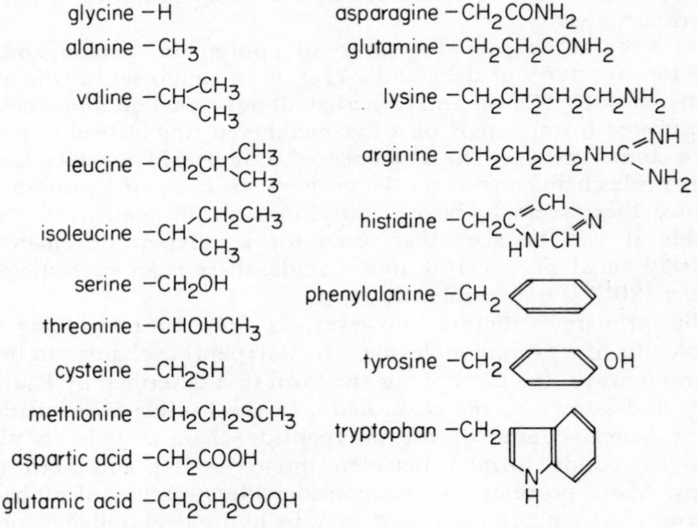

glycine	$-H$	asparagine	$-CH_2CONH_2$
alanine	$-CH_3$	glutamine	$-CH_2CH_2CONH_2$
valine	$-CH(CH_3)_2$	lysine	$-CH_2CH_2CH_2CH_2NH_2$
leucine	$-CH_2CH(CH_3)_2$	arginine	$-CH_2CH_2CH_2NHC(=NH)NH_2$
isoleucine	$-CH(CH_2CH_3)CH_3$	histidine	$-CH_2C{=}CH{-}N{=}CH{-}NH$ (ring)
serine	$-CH_2OH$	phenylalanine	$-CH_2C_6H_5$
threonine	$-CHOHCH_3$	tyrosine	$-CH_2C_6H_4OH$
cysteine	$-CH_2SH$	tryptophan	$-CH_2$-indole
methionine	$-CH_2CH_2SCH_3$		
aspartic acid	$-CH_2COOH$		
glutamic acid	$-CH_2CH_2COOH$		

FIGURE 9 The side chains (R of Fig. 8) of the 19 α-amino acids (excluding proline) occurring generally in proteins.

kind known as *hydrophobic interactions,* which may be roughly explained by saying that the more water-insoluble or 'oily' amino acid side-chains are more prone to lie in contact with each other than with the solvent water molecules. The formation of large protein molecules by hydrophobic aggregation of folded polypeptide chains is called *quaternary structure,* and it usually shows a rather precise symmetry in the arrangement of the sub-units.

Proteins are very accurately determined molecules. Although a virtually infinite variety is possible, any given species produces a finite though large number of kinds of protein, each one quite constant in its enormously complex structure. It might seem unlikely, at first sight, that there could exist in the cell any information store large enough to carry the specifications for all the proteins, unless this store were in the proteins themselves. The problem was, however, somewhat simplified by the demonstration that all the information for the secondary and higher-order structure of a protein is implicit in the primary structure, or amino acid sequence.

It has been shown for a number of proteins that, through treatment with substances such as urea which disrupt hydrogen bonds, and with reducing agents to break the disulphide bridges where these exist, the entire folding structure of the molecule can be undone. In several such cases the original structure is reformed spontaneously when the urea and reducing agents are removed. Observations of this kind imply that, given the information for the primary structure—the amino acid sequence—the various orders of folding structure, complex as they are, present no additional problem; they merely represent the most stable arrangement of the polypeptide chains under normal intracellular conditions.

Any one of the three classes of compounds we have been considering might seem to qualify, by reason of its structural complexity, as a suitable material for the storage of genetic information. Of the three, the proteins, being made out of twenty kinds of unit rather than from only four, might seem more promising material for a code than either DNA or RNA. So, indeed, it was thought when the problem of the genetic material was first posed in chemical terms. Present knowledge, however, leaves no doubt that the genetic material is DNA or, in some viruses, RNA, and that the information necessary for the accurate synthesis of the proteins of the cell is not carried in the proteins themselves but rather in the nucleic acids.

THE SELF-REPLICATION OF DNA

A property which we should expect to find in the genetic material is the capacity for self-replication, that is for directing the formation of more material like itself. No mechanism has ever been suggested through which proteins could self-replicate. In the case of DNA, on the other hand, not only is there a fairly obvious mechanism, but the mechanism has actually been proved correct by experiment.

Watson and Crick, who in 1953 first proposed the now generally accepted double-helical structure for DNA, were also the first to point out the possible relevance of this structure to the problem of self-replication. As we have seen, the fact that thymine always pairs with adenine and cytosine with guanine means that the sequence of bases along one strand is determined by the sequence along the other. Thus, if we imagine a double helix being separated by unwinding into its component strands A and B, and a new strand being synthesized

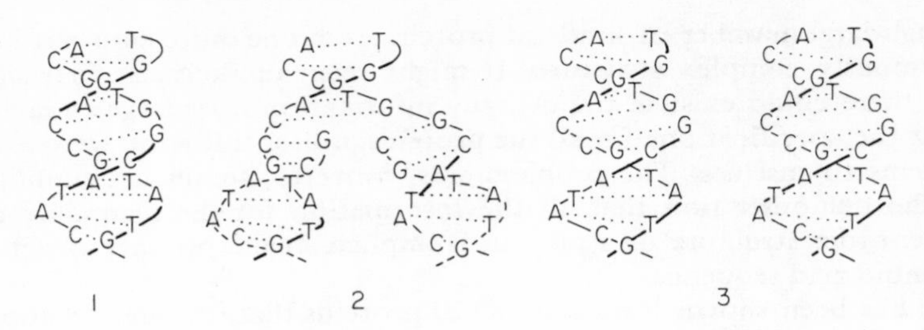

FIGURE 10 The Watson–Crick model for DNA replication, with unwinding of the two strands of the original double helix and the concomitant synthesis of a new complementary strand along each. The bases are shown in an imaginary sequence and are represented by their initial letters; 1, 2 and 3 represent successive stages.

along each old one, the new strand formed alongside A will necessarily be a replica of B, and the new strand formed alongside B must be a replica of A. Fig. 10 illustrates the idea. The result is two DNA double-stranded molecules each identical with the 'parent' molecule. This process is termed *semi-conservative* replication, since each new molecule contains one old strand and one new one.

There have been a number of experiments to prove the semi-conservative model, but the one which first really established the point was that of Meselson and Stahl, described in 1958. These authors grew the bacterium *Escherichia coli* for about 14 generations in a growth medium containing, as sole source of nitrogen, ammonium chloride with 96 per cent of its nitrogen in the form of the heavy isotope ^{15}N. As a result all the nitrogenous components of the cells, including the nucleic acids, were of higher than normal density.

It was shown that ^{14}N- and ^{15}N-containing DNA could be separated from each other by the method of *density gradient centrifugation.* In this procedure the sample to be analysed is added to a concentrated solution of caesium chloride in the cell of an analytical ultracentrifuge, which is then run at high speed for about 24 hours. Owing to the tendency of the relatively large caesium ions to sediment in the very high gravitational field a density gradient is set up, with the highest density at the bottom of the cell and the lowest at the top. The DNA comes to lie in a narrow band at a level corresponding to its own density, and its position can be recorded at any time photographically, making use of the very high optical absorption of nucleic acids in the ultraviolet wavelength range. The concentration of the caesium chloride is adjusted so that the DNA forms a band about half-way down the cell, and other cell constituents either float to the top or sink to the bottom. DNA containing ^{15}N forms a band slightly further down the gradient than DNA of normal density.

What kind of DNA will be found in cells which, originally fully 'labelled' with ^{15}N, are allowed to multiply for different periods of time in medium containing only ^{14}N? There were three main possibilities. Firstly, the pre-existing DNA might have been dispersed in small fragments among the progeny cells, or broken down completely, so that all the progeny DNA contained ^{14}N only. Secondly, the parental ^{15}N DNA might have been conserved and transmitted intact to progeny cells without separation of strands, in which case all the newly formed DNA would contain only ^{14}N but the ^{15}N-DNA would remain detectable until diluted out by growth. Finally, if the semi-conservative model of Watson and Crick were correct, all the DNA present after one round of cell division in ^{14}N medium should be of intermediate density, each molecule containing one old ^{15}N strand and one new ^{14}N strand. After two cell divisions the DNA should consist of a fifty-fifty

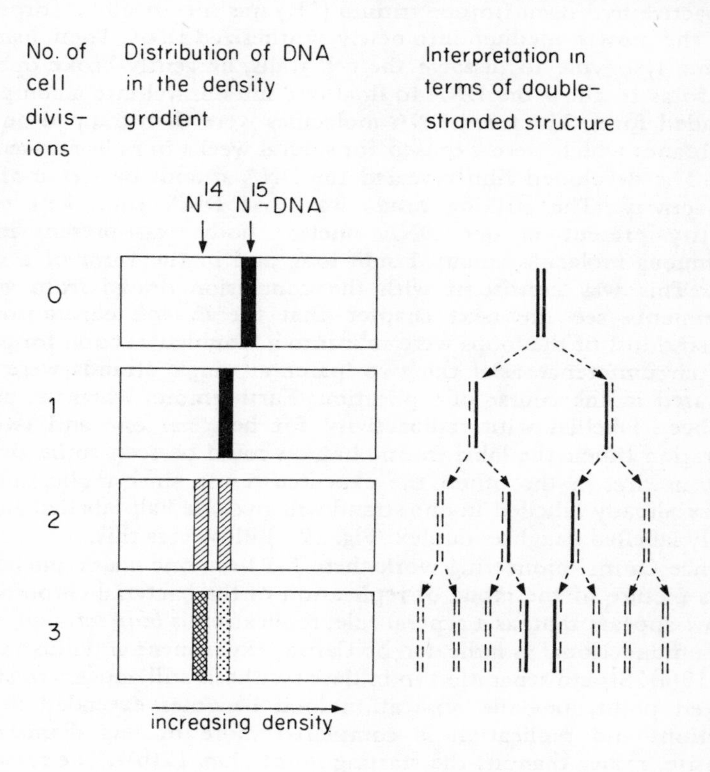

FIGURE 11 The Meselson–Stahl experiment (see text). The left half of the diagram represents what is actually seen in the ultracentrifuge photographs, with relative intensities of DNA bands indicated by different degrees of stippling. On the right is a diagrammatic interpretation of the results in terms of double-stranded structure and semi-conservative replication, with newly synthesized (light) strands shown as dotted lines.

mixture of wholly ^{14}N molecules and molecules of intermediate density, while after three cell divisions the same two kinds of DNA should be found but now in the ratio 3:1. The actual experimental results showed the correctness of the third alternative in a strikingly clear and convincing fashion (see Fig. 11).

The semi-conservative replication of DNA shown by the Meselson and Stahl experiment can most easily be imagined as occurring through the two strands of the 'parental' duplex molecule becoming separated, with new synthesis, adding a new complementary single strand to each of the separated strands, following closely after the strand separation. This mode of replication can actually be seen under the microscope thanks to an ingenious procedure pioneered by Cairns. He grew *Escherichia coli* for a period of time adequate for nearly two cell fissions to take place under conditions in which thymine containing the radioactive hydrogen isotope tritium (^3H) was specifically incorporated from the growth medium into newly synthesized DNA. Then, using the enzyme lysozyme to dissolve the cell walls, he gently broke open the cells so as to allow the DNA to float out and unravel into a completely extended form. The freed DNA molecules were then trapped on filter membranes which were exposed for several weeks to radiation-sensitive film. The developed film revealed the DNA strands by virtue of their radioactivity. The striking result was that DNA equivalent to the quantity present in one DNA nuclear body was present in one continuous molecule, about 1 mm long and in the form of a closed loop. This was consistent with the conclusion drawn from genetic experiments—see the next chapter—that the *E. coli* chromosome is circular. Most of the loops were split into a 'doubled' section for part of their circumference as if the two 'parental' single strands were being separated in the course of replication. Furthermore, when the culture had been labelled with radioactivity for between one and two cell generation times, the label in one branch could be seen to be twice as heavy as that in the other—the expected result, since replication of a duplex already labelled in one strand will give one half-labelled and one wholly labelled daughter duplex. Fig. 12(a) illustrates this.

Since Cairns' pioneering work there has been one major amendment to his picture of the mode of replication of the bacterial chromosome. It now appears that, as a general rule, replication is *bidirectional*, rather than unidirectional as indicated by Cairns' experiment and illustrated in Fig. 12(a). Strand separation to initiate synthesis still appears to start at a fixed point, but the replication loop becomes extended in both directions and replication is completed more or less diametrically opposite, rather than at, the starting point (Fig. 12(b)). The reason for the unidirectional mode apparently shown by Cairns' original observations is not clear—perhaps in some cells, for some reason, only one of the two potential replication forks is active.

We may note in passing the fantastically high degree of coiling and packing which must be necessary to get an extended length of about 1 mm of DNA into a cell no more than 1μm (i.e. 1/1000 mm) long.

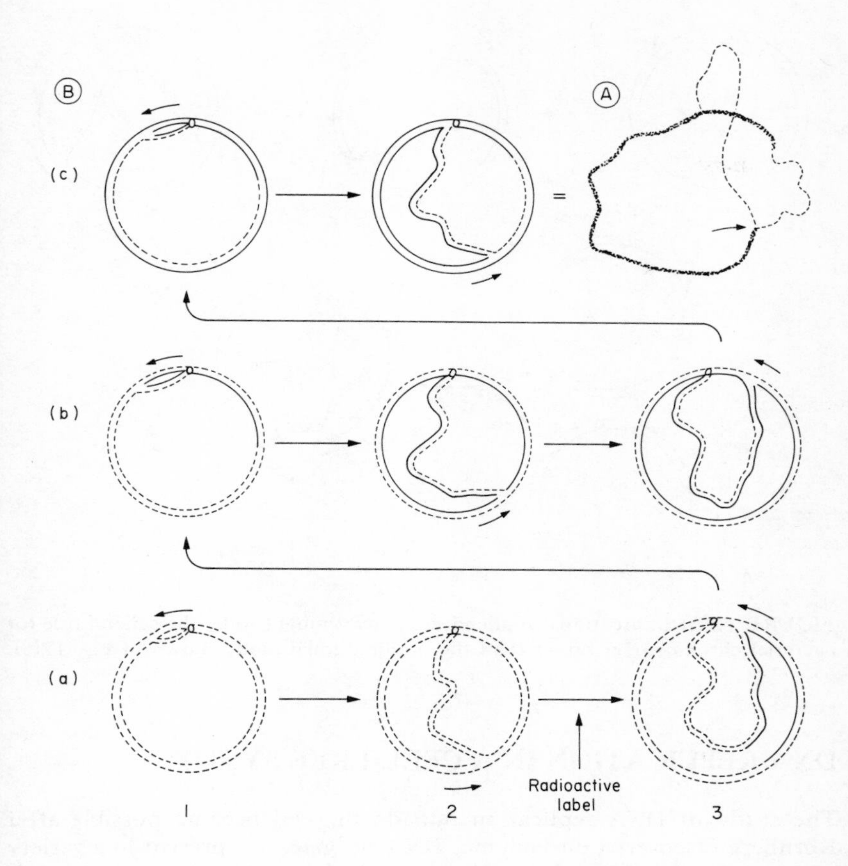

FIGURE 12(a) (A) The appearance of one of Cairns' autoradiographic pictures of a replicating *Escherichia coli* chromosome, with the different parts of the structure either singly (light stippling) or doubly labelled (heavy stippling).
(B) The explanation of how the observed pattern could have arisen by semi-conservative replication for two replication times following the introduction of tritiated thymidine as radioactive DNA precursor. 1, 2 and 3 represent respectively the beginning, middle and end of rounds of replication (a), (b) and (c). Unlabelled single strands of DNA are shown as dotted and labelled as full lines. The point at which replication begins and at which the two daughter DNA loops finally separate is shown as a small circle. The replication fork and its direction of travel is indicated by a small arrow. Note that the scheme depends on DNA replication going round the loop in one direction only (i.e. only one replication fork, the other end of the double section being a stationary swivel-point). Although indicated by Cairns' data, this may be a special case, since bi-directional replication (with two replicating forks travelling away from one another and meeting at the finishing point) has now been shown to occur in *E. coli.*

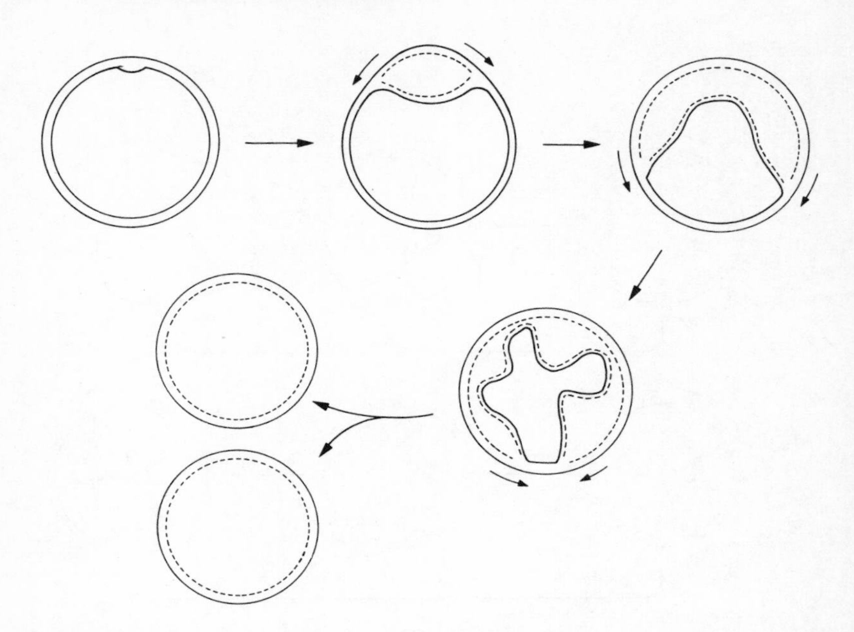

FIGURE 12(b) Bidirectional replication – now thought to be the general rule for bacterial chromosomes rather than the unidirectional mode shown in Fig. 12(a).

DNA REPLICATION IN A CELL-FREE SYSTEM

The study of DNA replication outside the cell became possible after Kornberg discovered an enzyme, *DNA polymerase,* present in a variety of bacteria and cells of other kinds, which would synthesise DNA if given all four deoxynucleoside triphosphates, magnesium ions, and a small amount of initial DNA to act as 'template'. The 'template' could be either single- or double-stranded DNA, although single-stranded DNA worked better. The DNA synthesised in the system resembled both in its base composition and (so far as could be determined) in its base sequence the DNA used as the template when the latter was double-stranded. When single-stranded template was used the synthesized DNA, at least after short times of incubation, resembled that expected if it were complementary to the template strand (Table 1).

A most spectacular and convincing demonstration of cell-free DNA synthesis was achieved by Goulian, Kornberg and Sinsheimer who succeeded in showing that the infectious DNA of a simple bacteriophage could be reproduced with the aid of DNA polymerase in a test-tube system. The DNA of the simple phage ϕX174 occurs in single-stranded form in the virus particles, but it is converted to a double-stranded form in the course of its multiplication in the infected bacterial cell. Goulian and his colleagues used the single-stranded isolated DNA from infective virus particles (*plus* strands) as a template for synthesis of the complementary (*minus*) strand. This latter was, by a special technical trick which we need not describe, separated from the original plus strand and used, in its turn, as a template for synthesis of more plus strands. The synthetic plus strands, made entirely by the enzyme in the cell-free system, were shown to be fully capable of initiating virus multiplication in *E. coli.* The infectivity of isolated bacteriophage DNA will be mentioned again later in this chapter as one proof that DNA is indeed the genetic material.

TABLE 1

Synthesis of DNA catalysed by a cell-free enzyme and governed by the base composition of added DNA template

The reaction mixtures contained the four deoxyribonucleoside triphosphates, purified enzyme from *Escherichia coli,* magnesium ions and DNA as template, and were buffered with phosphate. The nucleoside triphosphates were labelled with radioactive carbon (^{14}C), and the base composition of the synthesised DNA was measured by determining the distribution of radioactivity among the four bases of the DNA present at the end of the reaction period. Data of Kornberg et al. (1958). *Proc. Nat. Acad. Sci., Wash.,* 44, 1191.

Source of DNA template		Molar proportions of bases			
		Adenine	**Thymine**	**Guanine**	**Cytosine***
Mycobacterium phlei	Template	0.65	0.66	1.35	1.34
	Product	0.66	0.80	1.17	1.34
Aerobacter aerogenes	Template	0.90	0.90	1.10	1.10
	Product	1.02	1.00	0.97	1.01
Escherichia coli	Template	1.00	0.97	0.98	1.05
	Product	1.04	1.00	0.97	0.98
Calf thymus	Template	1.14	1.05	0.90	0.85
	Product	1.19	1.19	0.81	0.83
T2 bacteriophage	Template	1.31	1.32	0.67	0.70*
	Product	1.33	1.29	0.70	0.70*

*Hydroxymethylcytosine in T2

Paradoxically, after all this elegant and convincing work, it seems that the DNA polymerase isolated by Kornberg (now called polymerase I) is not the enzyme responsible for the major DNA replication which occurs during bacterial cell growth and division. Cairns has succeeded in isolating genetic variants (mutants) of *E. coli* which lack the enzyme; these can *still grow* at a near normal rate, though they are abnormally sensitive to ultraviolet light. The Kornberg enzyme probably has a normal function in *repair* of DNA following radiation damage rather than of major new synthesis. The latter function seems to be carried out by another enzyme (polymerase III) with many of the same properties but more difficult to isolate because it is bound firmly to cell membranes. Mutant strains producing unusually heat-sensitive polymerase III have been isolated and these neither synthesize DNA nor grow at high temperatures (42°C) at which normal *E. coli* DNA synthesis and growth are unimpaired.

GENETIC TRANSFORMATION WITH DNA

If DNA is the genetic material it must not only be able to direct its own replication; it must also provide the ultimate control over everything else that happens in the cell. This may seem an ambitious claim to make for any substance, but it is a claim that has been justified, at least for a number of species of bacteria, in a most clear and direct way.

In *Diplococcus pneumoniae* (or 'pneumococcus'), *Bacillus subtilis,* several species of *Hemophilus* and several other species of bacteria, it is possible to take any two genetically distinct strains of the same species and transfer the differentiating genetic traits from strain to strain by incubating cells of the one with purified DNA isolated from the other. This phenomenon, known as *transformation,* was first discovered in pneumococcus by Griffith, and the identification of the transforming principle as DNA was achieved by Avery, McLeod and McCarty in 1944. The genetic trait with respect to which transformation was first observed was the ability to produce a polysaccharide cell capsule of a particular type; the presence and specific nature of this capsule is important in determining the serological properties and also the pathogenicity of the pneumococcus. Different strains produce capsules of different types, and some strains long-cultured in the laboratory lose the ability to produce a capsule of any kind. Cells of a capsule-deficient strain can acquire, with a certain probability which may be several per cent under the right conditions, the ability to make a capsule of a specific type as a result of being incubated in the presence of DNA purified from cells of that particular type.

Transformation is, however, most conveniently studied with respect to genetic differences which determine ability versus inability to grow on certain media. With a system of this sort the frequency of transformation from inability to ability to grow is easily measured by spreading known numbers of treated cells on the differentiating medium and counting the number which grow to form colonies.

Fig. 13 illustrates the results of an experiment in which cells of a strain of *B. subtilis,* unable to synthesize the essential amino acid tryptophan and thus unable to grow on tryptophan-free medium, were treated with DNA from another strain which could make its own tryptophan. The number of transformed cells, as revealed by the number of colonies appearing on tryptophan-free medium, was proportional to the concentration of DNA over a wide range of concentrations. At very high DNA concentrations the yield of transformed cells (transformants) was evidently limited by other factors. It will be seen that even a concentration of DNA of one nanogram per cm^3 (one part per thousand million) is sufficient to bring about considerable numbers of transformations. In fact, this kind of experiment does not reveal the full potency of the DNA, since not all the DNA present in the incubation mixture is necessarily taken up by the bacterial cells. The best way of determining how much of the DNA is taken up is to isolate it from bacteria which have been grown on medium containing the radioactive phosphorus isotope (^{32}P) in its phosphate. The DNA thus obtained is 'labelled' with radioactivity, and the amount taken up by the treated bacteria can quite easily be determined by measurements of the amount of radioactivity which cannot be removed from them by washing.

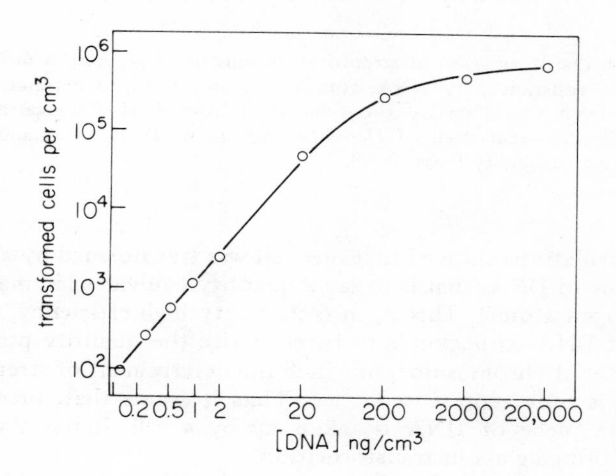

FIGURE 13 Transformation of tryptophan-dependent *Bacillus subtilis* to tryptophan-independence, by DNA from a tryptophan independent strain, as a function of DNA concentration in the medium. Data from J. Spizizen, *Proc. Nat. Acad. Sci., Wash.*, 44, 1072.

Fig. 14 shows the results of an experiment on *Hemophilus influenzae*, in which a streptomycin-sensitive strain was treated with ^{32}P-labelled DNA from a streptomycin-resistant strain. The numbers of transformed cells were determined by counting the numbers of colonies formed when samples of the treated bacterial suspensions were spread on medium containing enough streptomycin to kill all untransformed cells. The number of transformants was nearly proportional to the amount of DNA taken up by the suspension.

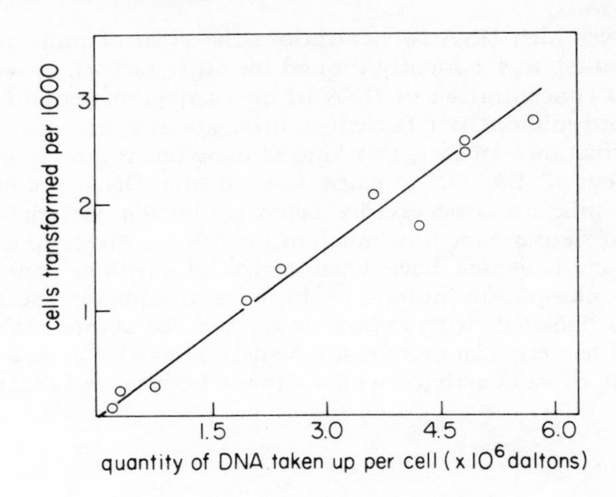

FIGURE 14 Transformation of streptomycin-sensitive *Hemophilus influenzae* to streptomycin resistance, by DNA from a resistant strain, as a function of the amount of DNA taken up by the cells. Data from S. H. Goodgal and R. M. Herriott, *The Chemical Basis of Heredity,* editors W. D. McElroy and B. Glass, Johns Hopkins University Press, 1958.

The calculations showed that one cell was transformed by about 2×10^9 daltons of DNA (that is to say a quantity equivalent in mass to 2×10^9 hydrogen atoms). This is, in fact, a very high efficiency, since this amount of DNA corresponds to barely twice the quantity present in a single bacterial chromosome, in which the determinant of streptomycin resistance is only represented once. Thus it seems that, provided the appropriate piece of DNA is taken up by a cell, it has a very high chance of bringing about transformation.

It seems clear from the direct proportionality between the rate of transformation and the amount of DNA absorbed, which extends down to very low DNA concentrations, that *one* piece of DNA can transform a cell with respect to any given trait, and that, once transformed, this cell can transmit the new trait to an indefinitely large number of descendants. The descendants can, in fact, be used in their turn as a source of DNA for inducing the same transformation in other cells.

Within species of bacteria for which transformation can be demonstrated it seems that any stable genetic trait can be transferred experimentally from a strain possessing it to another lacking it through the agency of DNA. Transformation with respect to a given character difference can occur in either direction. For instance, in the example from *Hemophilus influenzae* just mentioned, it would have been possible to transform the streptomycin-resistant strain to streptomycin sensitivity, though, of course, the transformants would have been technically more troublesome to isolate in this case.

Transfer of traits even between different species has been demonstrated in the genus *Hemophilus,* although the efficiency of transformation is less when the species are more distantly related. However, even within species where transformation can regularly be demonstrated, cells are not transformable at all times. Pneumococcus cells, for example, go through phases of alternating high and low transformability which are related to the cell division cycle. In *B. subtilis* the cells are most easily transformable when the culture is approaching saturation and cell division has almost ceased.

The demonstration of transformation in any bacterial species is probably just a matter of finding the right conditions. The reason it remained for long undemonstrated in *Escherichia coli* appears to be that DNA taken up by cells of this species is normally rapidly destroyed by hydrolytic enzymes (nucleases). It has now been shown that when suitable nuclease-deficient mutant strains of *E. coli* are used as recipients transformation does, in fact, work very well.

Considerable evidence is now available on the molecular mechanism of transformation, the detail of which is beyond the scope of this book. It must suffice to note that, while effective preparations of DNA must consist of *double-stranded* molecules if they are to be effective in transformation, there is good evidence that, after uptake, only *one* strand of any given section of a DNA molecule is actually integrated into the chromosome of the recipient cell. Single-strand integration gives initially a hybrid duplex molecule in which the two strands may carry different donor and recipient genetic determinants. Such hybridity is eliminated following the next round of DNA replication (if not corrected before) giving one daughter duplex of donor and one of recipient type. Genetic transformation illustrates the principle, which we shall meet again at the end of this chapter in connection with viruses, that a *single* nucleic acid strand suffices for transfer of information from one cell to another.

THE GENETIC MATERIAL OF VIRUSES

Viruses do not in themselves possess the characteristics of living organisms. In isolation, a virus particle is incapable of any kind of metabolic activity or growth. When it gains access to a host cell, however, an infective virus can bring about the synthesis of many more virus particles of the same kind. Evidently the viral material which enters the host cell is able to turn the metabolic apparatus of the host from its normal ends to the task of producing virus. Thus a virus infection can be regarded as the capture of the control of the cell by foreign genetic material.

In view of the known role of DNA as the genetic material of bacteria, it is not surprising that the infective material of some viruses is also composed of DNA. This was first shown to be true for the bacterial virus (bacteriophage) T2, which attacks *Escherichia coli.* T2, like its close relative T4, is a relatively large virus with a complex structure consisting of a head, which is in effect a protein capsule with a filling of DNA, and a tail through which the virus makes attachment to the surface of the bacterium. Hershey and Chase, by first labelling the DNA of the virus with radioactive phosphorus (^{32}P), and then the protein with radioactive sulphur (^{35}S), were able to show that only the DNA from the virus actually entered the host cell. The empty protein heads remained outside, and could be removed from the cell surfaces by mechanical agitation without disinfecting the bacteria.

The phage tail serves as an injection tube through which the DNA passes into the host. Once inside the bacterium the phage DNA causes all normal bacterial growth to stop; the whole metabolic apparatus of the cell now becomes directed towards the production of specifically *viral* DNA and proteins. Some of the latter are used in making new phage heads and tails, while others are enzymes functioning in phage DNA synthesis or perform functions in phage assembly without actually becoming incorporated into the completed virus particles. During the formation of the phage components special kinds of RNA also appear in the infected cell; these have a 'messenger' function, carrying programmed instructions from the phage DNA for the synthesis of phage proteins (see Chapter 5). Finally, about 20 or 30 minutes after infection, the bacteria burst (undergo *lysis*) and about a hundred mature bacteriophage particles are liberated from each cell. All these malign and complicated processes are the consequence of the injection into the cell of DNA of a specific type.

Since the genetic role of DNA was demonstrated in the T-even phages it has been shown that infection of *Escherichia coli* with several other bacteriophages (such as *lambda* and ϕX174) can be accomplished using isolated phage DNA alone, without participation of virus protein. Such infection requires the prior removal, by enzymic digestion or otherwise, of the bacterial cell wall, leaving a naked *protoplast*, through the membrane of which DNA can penetrate without the use of the normal virus injection apparatus.[1]

The hypothesis that *all* virus genetic material is DNA cannot be held for the simple reason that not all viruses contain DNA. A good many of the smaller and structurally comparatively simple viruses have RNA instead. The best studied of these is *tobacco mosaic virus* (TMV), which has perhaps the simplest possible structure for a virus. Identical sub-units, each one consisting of a single polypeptide chain, are packed in a helical fashion to form a hollow protein rod, down the centre of which is wound a single strand of RNA. Fraenkel-Conrat was able to separate the protein from the RNA under mild conditions, and then to cause the two components to reassociate to form rods which resembled the original virus though with somewhat reduced infectivity.

When two different strains (differing in the symptoms induced in the host tobacco plant) were each dissociated, and the RNA of one

[1] A process called *transfection*.

reassociated with the protein of the other, and *vice versa*, the type of virus which was propagated when the resulting 'hybrid' particles were used to infect a host plant was always that from which the RNA had been derived. It was later found that the RNA in isolation was somewhat infective, though it loses its infectivity much more easily than does the intact virus. The protein alone cannot infect. It is thus clear that RNA is the genetic material of tobacco mosaic virus. Similar conclusions have been drawn from studies of a number of animal viruses containing RNA but no DNA, including polio-virus, encephalitis virus and several others. In a number of these animal viruses infectivity of isolated RNA has been demonstrated.

We must therefore admit that both DNA and RNA can act as genetic material, though so far RNA has been shown to have this function only in certain viruses. In fact there are strong suggestions that RNA and DNA may act as alternate carriers of genetic information for the same virus. Some animal viruses, with only RNA and no DNA in the infective particle, determine the formation of a *reverse transcriptase* which can catalyse the synthesis of DNA on a RNA template. Thus the information for virus development may be transcribed after infection into DNA, and be maintained as such during a prolonged latent phase in the host cell.

We have seen that there is a well-founded theory of DNA replication based on its double-stranded structure. However, since the two strands are complementary, and the sequence of one implies that of the other, either one of them is in principle sufficient for information transfer during virus infection. Thus it is not altogether surprising to learn that many viruses have single rather than double-stranded nucleic acid in their infective particles. It so happens that most known DNA viruses have double-stranded DNA and most known RNA viruses single-stranded RNA, but several examples of single-stranded DNA viruses (e.g. ϕX174 mentioned on p. 20) and of double-stranded RNA viruses are known. The single-stranded viruses replicate their nucleic acid in the infected cell via a double-stranded intermediate and so their existence does not imply a unique mode of self-replication.

The replication of the RNA of RNA viruses requires the presence of an enzyme analogous to DNA polymerase but utilising RNA instead of DNA as a template and ribonucleoside rather than deoxyribonucleoside triphosphates as substrates. Such enzymes (sometimes called *RNA replicases*) have been purified from bacteria infected with RNA bacteriophages; they are determined by the viral genetic material and form no part of the normal equipment of the host cell.

Thus, taking virus as well as normal cellular systems into account, we see that the storage and mode of transmission of genetic information can take quite varied forms. The basis of the system, however, is always a nucleic acid owing its specificity to its unique base sequence and its capacity for accurate replication to the principle of specific hydrogen-bonded base pairing between the two complementary strands of a duplex.

3 Mapping the Genome

INTRODUCTION

We have seen that the genetic material consists universally of DNA or (in certain viruses) RNA. Since DNA molecules are always extremely large, with only one molecule per nucleus in prokaryotes or, as is now known, only one per chromosome in at least simple eukaryotes, it is clear that they must be divided into many functionally different segments (*genes*) controlling different things. The complete ensemble of genes is called the *genome.*

Until recently the possibilities for analysing the internal arrangement of the genome by physical and chemical methods have been negligible. During the last few years, impressive advances have been made in physical genetic mapping, and these are described in Chapter 8. However, the information obtainable by these new methods is still very limited and confined to very simple genomes. The great amount of knowledge which we now have about the arrangement of genes in organisms of all kinds is still due overwhelmingly to the application of specifically genetical methods, in which deductions are made from the patterns of hereditary transmission shown by the genetic determinants of visible characteristics. Strictly genetical mapping depends on exploiting the natural history of the organism under study rather than on physics or chemistry. This chapter describes some of the most important mapping techniques used for different organisms and for viruses.

NEUROSPORA–AN ORGANISM WITH A REGULAR SEXUAL CYCLE

As was explained in Chapter 1, *Neurospora crassa* has a sexual reproductive cycle similar in essentials to that of higher organisms. There is a fusion of equal haploid nuclei in the young ascus, and an orthodox meiosis in the maturing ascus leading to the formation of the haploid ascospores.

As a result of meiosis homologous segments of genetic material, originally contributed by different parents, are necessarily *segregated* from each other. Where one chromosome of a homologous pair in the diploid ascus nucleus carries one variety of a particular gene, and the other carries a different variety, two of the ascospore pairs will receive one gene and two the other.

This one-to-one segregation of unit genetic differences among the products of meiosis is found to be almost invariably the rule in *N. crassa* and other organisms with a regular sexual cycle. The inheritance of mating type is the most obvious example in *N. crassa*. The parents of any cross necessarily differ in mating type, and each ascus contains four spores (representing two products of meiosis) which germinate to give cultures of one mating type, and four spores which give cultures of the other mating type. The property of mating type serves, in fact, as a *marker* through which the transmission of the chromosome segment determining it can be followed.

Now what happens when there is a second marker differentiating the parental strains? To take an example, suppose the strain of mating type *A* has a compact 'colonial' type of growth habit (symbolised by *col*), while the *a* parent is like wild type in its growth (col^+). Both the *A: a* difference and the *col*: col^+ difference show regular one-to-one segregation in each ascus but the two markers segregate quite independently of each other. In other words each *A* spore has equal chances of being *col* or col^+, and the same applies to each *a* spore. Among all the ascospores produced by the cross, the four types *A col*, *A* col^+, *a col* and *a* col^+ occur with equal frequencies. The explanation of this independent behaviour of the two markers is that they are associated with two different members of the haploid chromosome set. The way in which the two chromosomes of one pair segregate from each other during meiosis is usually quite independent of any other pair.

There are many hundreds of unit differences which provide genetic markers in *N. crassa*, so each of the seven chromosomes must carry many such markers. If chromosomes were inherited as indivisible units two markers on the same chromosome would have to be completely linked in their inheritance. For example, if the *A* parent was albino (*al*) while the *a* parent had the normal orange-pink pigmentation (al^+), and if mating type and albinism were controlled by different segments of the same chromosome (as is known to be the case), the simplest expectation would be that all the ascospores formed in the cross would be either *A al* or *a* al^+. What is actually found is quite different. The result does indeed differ from the case of independent segregation, but considerable numbers of *recombinant* spores of the constitutions *A* al^+ and *a al* nevertheless occur. Very roughly the frequencies of ascospore types found in this instance are: *A al* 30%, *a* al^+, 30%, *A* al^+ 20%, *a al*, 20%. It is characteristic of this kind of result that the two parental types are about equal in frequency, as are the two recombinant types. A convenient inverse measure of the degree of linkage shown by the two markers is the percentage of recombinants among the meiotic products, in this case 40 per cent. For different pairs of linked markers this percentage can vary from near zero, for very tight linkage, to almost 50 per cent for very loose linkage.

All the known markers in *N. crassa* can be classified into seven *linkage groups*. Members of different groups always segregate independently, while members of the same group nearly always show some degree of linkage (less than 50 per cent recombination) and even where they do not their relationship is revealed by their common linkage to other markers in the same group.

It is possible to represent each linkage group by a linear map, the distance separating any two markers being related to the amount of

TABLE 2
Analysis of a Neurospora cross segregating with respect to four linked markers (data of Strickland, *Genetics*, 46, 1125 (1961)).

Parental genotypes (linkage group 5)	*hist*	+	*bis*	+	x	+	*inos*	+	*paba*

Chromosome pair at diplotene*	*hist*	+	*bis*	+
	hist	+	*bis*	+
	(a)	(b)	(c)	
	+	*inos*	+	*paba*
	+	*inos*	+	*paba*

No. of asci	Constitutions of ascospore pairs	Interpretation
1597	*hist* + *bis* + *hist* + *bis* + + *inos* + *paba* + *inos* + *paba*	No exchange in *hist–paba* region
282	*hist* + *bis* + *hist inos* + *paba* + + *bis* + + *inos* + *paba*	Exchange between *hist* and *inos*–(a) above
329	*hist* + *bis* + *hist* + + *paba* + *inos bis* + + *inos* + *paba*	Exchange between *inos* and *bis*–(b) above
564	*hist* + *bis* + *hist* + *bis paba* + *inos* + + + *inos* + *paba*	Exchange between *bis* and *paba*–(c) above
38	13 other types	Different combinations of double exchanges

Symbols: *hist, inos, paba*–growth requirement for histidine, inositol and p-aminobenzoic acid respecitvely; *bis*–'biscuit' (distinctive growth habit).

*cf. Fig. 2.

recombination which they show. For fairly low frequencies of recombination (i.e. less than 20 per cent or so) the frequencies are nearly additive. That is to say, if the map order of three markers is A—B—C, the frequency of recombination shown by A and C is close to the sum of the frequencies shown by A and B and by B and C. The conventional unit of map distance is one per cent recombination.

There is no doubt that the seven linkage groups correspond to the seven chromosomes of the haploid set. Recombination within the same linkage group is nicely accounted for if the chiasmata visible at diplotene are due to exchanges of homologous chromatid segments, by breaking and rejoining or some equivalent process. Every such exchange, or cross-over, gives two recombinant chromatids leaving two of parental constitution, so the percentage recombination shown by two linked markers would be equal to one half of the percentage incidence of chiasma formation between them if there were never more than one chiasma per chromosome pair. Actually, multiple exchanges can occur between distant markers, and map distances can exceed 50 units. Recombination percentages, however, seldom exceed 50 since, so far as recombination is concerned, the occurrence of a second exchange is as likely as not to cancel out the effect of the first. This is why recombination frequency is only an accurate measure of map distance when the interval concerned is short enough for multiple exchanges to be negligible.

Because cross-over frequencies tend to vary somewhat from one cross to another, the order of three linked markers is best determined from a cross in which all three are segregating simultaneously. Analysis of ascospores from a *three-point cross* usually fixes a best order for the three markers such that the common recombinant types can be accounted for by single exchanges. Except where the markers are very widely spaced recombinant types which have to be attributed to double exchanges are much less frequent. Table 2 illustrates the principle.

Exactly the same principles of mapping apply to yeast, although here linkages are harder to find because of the larger number of linkage groups (18 at the latest count), each one representing a different chromosome.

CONJUGATION AND RECOMBINATION IN *ESCHERICHIA COLI*

Bacteria do not show the regular alternation of haploid and diploid phases characteristic of neurospora and higher organisms. A variety of mechanisms exist in bacteria whereby homologous genes from different sources can be brought together in the same cell, but in none of these processes is a completely diploid cell regularly formed. Instead, a part of a genome, often little more than a fragment, can sometimes be transferred from a cell of one strain into the cell of another. The resulting 'zygote' is thus diploid for only part of its genetic material. Reduction to the normal haploid state, with genetic segregation and the possibility of recombination of linked markers, usually occurs within one or a few cell divisions following the formation of the partial diploid, but the details of the process through which this occurs are obscure; probably the extra fragment is simply lost through inability to replicate itself. There are three known ways in which this sort of transient partial diploidy can be achieved. These are cell conjugation, trasduction by bacteriophage, and transformation by isolated DNA. We will deal first with conjugation, which has been studied in great detail in *E. coli.*

The work of Lederberg and his colleagues in 1948 and the years immediately following showed clearly that strains of *E. coli* carrying different genetic markers could produce recombinants when mixed together. Lederberg was able to demonstrate that some markers were strongly linked, and he made considerable progress in constructing a linkage map. There were at first a number of inconsistencies in the map, and these were only cleared up following the findings of Hayes, in London, and Jacob and Wollman in Paris, on the peculiar genetic system of *E. coli.*

Not all *E. coli* strains can mate to give recombinants. Hayes was the first to recognise that sexual fertility depended on a special genetic factor (F), present in some strains and not in others. The most fertile crosses, in terms of frequency of recombinants produced, were between F^+ strains, possessing the factor, and F^- strains lacking it. Somewhat later a special kind of F^+ strain was discovered with was some hundred times more fertile in crosses with F^- than the F^+'s previously known. This was called Hfr, for *high-frequency recombination.* Hfr strains arise spontaneously, though with low frequency, in F^+ populations, and it seems that the fertility, such as it is, of F^+ strains is due at least partly, though perhaps not wholly, to the Hfr cells which they contain.

When F^+ and F^- cells are mixed together they form conjugating pairs with high efficiency. This pairing is dependent on the presence, on F^+ and Hfr but not on F^- cells, of distinctive filamentous appendages called *sex pili.* The functions of these pili are not fully understood. It is certain that they act at least as 'grappling irons', fastening at their tips to the surface of F^- cells and holding them in contact with their F^+ or Hfr partners. It has also been suggested that the transfer from cell to cell of genetic DNA takes place through the pili, which are probably hollow protein tubes of sufficient bore for the threading of a DNA double helix. This is not yet proved, though the DNA must, of course, be transferred through a connection of some kind if genetic recombinants are to be formed.

We will now consider an experiment carried out by Hayes, Wollman and Jacob which illustrates the main characteristics of the transfer process leading to recombinant formation. An Hfr strain which was essentially a wild type was able to synthesise all its own organic constituents including amino acids, was capable of utilising the sugars galactose and lactose, was susceptible to attack by bacteriophage T1, and was inhibited and killed by quite low concentrations of sodium azide or streptomycin. An F^- strain, in contrast, was unable to synthesise the amino acids leucine (*leu*$^-$) and threonine (*thr*$^-$), unable to utilise galactose (*gal*$^-$) or lactose (*lac*$^-$), resistant to T1 (*ton*^{r}), and also resistant to azide (*azi*^{r}) and to streptomycin (*str*^{r}) at concentrations which would kill the wild type organism. The contrasting wild type traits are symbolised by *thr*$^+$ *leu*$^+$ *gal*$^+$ *lac*$^+$ *ton*^{s} *azi*^{s} *str*^{s}.

Cells of the two strains were mixed together in suspension and, following an incubation period, the mixture was analysed for recombinants. In analysing any bacterial cross it is customary, in order to save labour, to use some automatic method for *selecting* recombinants, which may constitute only a very small minority of the cells present. In this instance recombinants were selected by spreading samples of the cell mixture on to solidified (agar) medium containing only the minimal nutrients essential for wild type growth (i.e. inorganic salts and glucose) and an inhibitory concentration of streptomycin. Such a medium will not support the growth of either parental strain and will select recombinants of the constitution *thr*$^+$ *leu*$^+$ *str*^{r}. The constitution with respect to the markers, *azi, ton, lac* and *gal* will not affect growth, and these may be termed *unselective* markers under the conditions of the experiment. When the constitution of the selected recombinants with respect to the unselective markers was investigated the result shown in Table 3 was obtained.

TABLE 3
Analysis of the cross F^- *thr*$^-$ *leu*$^-$ *azi*^{r} *ton*^{r} *lac*$^-$ *gal*$^-$ *str*^{r} × *Hfr thr*$^+$ *leu*$^+$ *azi*^{s} *ton*^{s} *lac*$^+$ *gal*$^+$ *str*^{s}

		Unselective marker			
		azi	*ton*	*lac*	*gal*
Percentage *thr*$^+$ *leu*$^+$ *str*^{r} recombinants with marker from:	Hfr parent	90	70	40	25
	F^- parent	10	30	60	75

It is apparent that the unselective markers are not being distributed randomly with respect to the selective markers. One could describe the situation by saying that *azi*^{s} and *ton*^{s} are linked, in different degrees, to *thr*$^+$ and *leu*$^+$, while *lac*$^-$ *and gal*$^-$ are linked to *str*^{r}. It may be noted in passing that *thr* and *leu* are actually separable markers, but as they are closely linked it was convenient to treat them as one unit.

There is one crucial feature of Hfr × F^- crosses which is not brought out by the data of Table 3. This is that Hfr cells are usually only able to transfer a part of their genome with high frequency to F^- cells. For example, Hayes' original strain HfrH, used in the cross under discussion, almost never transmits to recombinants a whole group of markers including genes governing ability to utilise maltose (*mal*) and mannitol (*mtl*) as carbon sources. Recombinant progeny practically always resemble the F^- parent with respect to these markers. This, together with other observations, led Hayes to suggest that the Hfr member of a conjugating pair of cells acted as a genetic donor, and the F^- as the recipient, and that the Hfr tended to donate only a part of its genome.

This hypothesis was strikingly confirmed by Jacob and Wollman who carried out the experiment of prematurely separating the conjugating pairs by subjecting the mating mixture to violent agitation after different periods of mating. They showed that the shorter the period of mating the fewer the Hfr markers which were able to appear among recombinants (Fig. 15). Eight minutes of undisturbed conjugation sufficed for the formation of some thr^+ leu^+ str^r recombinants, but when mating was interrupted after this time all the unselective markers originated from the F^- parent and none from the Hfr. After 9 minutes of uninterrupted mating the azi^s marker from the Hfr parent began to make its appearance among thr^+ leu^+ str^r recombinants, while the Hfr markers ton^s, lac^- and gal^- only began to appear after 11, 18 and 25 minutes respectively (Fig. 15).

These results are most easily interpreted as indicating a time sequence of injection of markers from the Hfr into the F^- cells, with *thr*

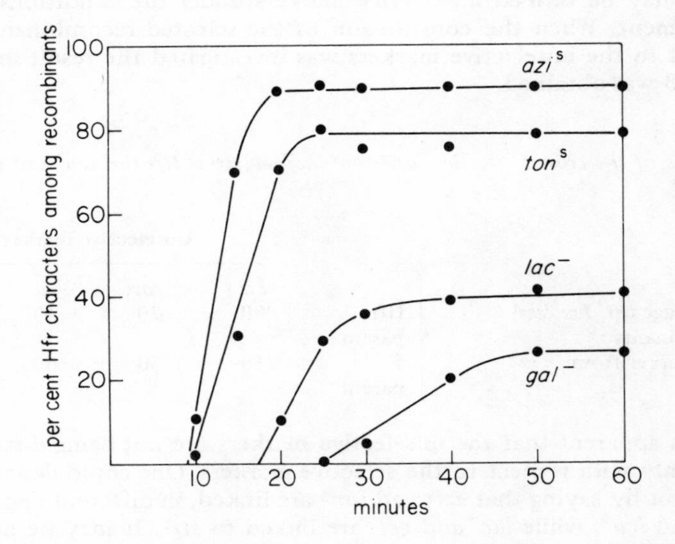

FIGURE 15 Results of interruption of mating in the *Escherichia coli* cross F— thr^- leu^- azi^r ton^r lac^- gal^- str^r × Hfr thr^+ leu^+ azi^s ton^s lac^+ gal^+ str^s. Conjugation was interrupted at various times after mixing by mechanical agitation, the cells were plated on streptomycin medium devoid of amino acids to select thr^+ leu^+ str^r recombinants, and the frequencies of the unselective Hfr markers among these recombinants were determined. Note that the markers show different maximum levels of incorporation as well as different times of entry. From Jacob and Wollman (1961).

and *leu* getting in relatively early followed, in order, by *azi, ton, lac* and *gal.* The interrupted mating technique furnishes, in fact, the most convenient method for mapping markers in a linear order in *E. coli.*

Further analysis of crosses of the $F^- \times$ Hfr type revealed an even more unconventional genetic feature. A considerable number of Hfr strains, other than HfrH which was the one used in the experiments detailed above, have been isolated at different times from F^+ cultures. These strains differ strikingly from HfrH, and from one another, in the particular block of markers which each is capable of transferring to F^- cells, and also in the time sequence in which the markers are injected. Just as HfrH injects the marker sequence *thr leu azi ton lac gal* (with some other markers falling within the same series), so each of the other Hfr's has its own specific sequence.

When the results from all available Hfr strains are collated it is seen that the sequences overlap in many cases, and that when they do so the

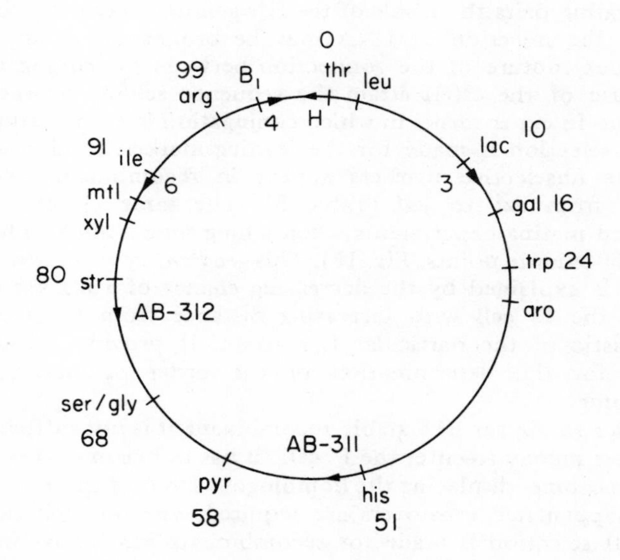

FIGURE 16 The circular linkage map of *Escherichia coli* derived from interrupted mating experiments using different Hfr strains. The arrows on the circle indicate the leading end and direction of entry of the linear chromosome injected by each of the Hfr strains, the designations of which are shown inside the circle. The figures round the outside of the map show distances, in minutes, based on time of entry experiments: The markers shown on the map are as follows: *thr*, threonine requirement; *leu*, leucine requirement; *lac,* lactose utilisation; *gal,* galactose utilisation; *trp,* tryptophan requirement; *aro*, requirement for shikimic acid or mixture of aromatic compounds; *his,* histidine requirement; *pyr,* pyrimidine requirement; *ser/gly*, requirement for serine or glycine; *str*, streptomycin resistance; *xyl,* xylose utilisation; *mtl,* mannitol utilisation; *ile,* isoleucine requirement; *arg,* arginine requirement. Simplified after Jacob and Wollman (1961).

order and spacing of the common markers is the same in different Hfr's, except that in some the order is *reversed* with respect to time of injection. Furthermore, and most unexpected of all, the composite map constructed from the marker time sequences of all the Hfr strains is a closed loop, or 'circle' (Fig. 16). This led to the conclusion that the *E. coli* genome (i.e. complete set of genetic material) consists of a closed loop of DNA in the F^- or non-Hfr cell, and that in Hfr cells the loop is broken prior to genetic transfer at a point which is characteristic of the Hfr strain, and injected as a linear structure. For the selection of recombinants from crosses involving different Hfr strains different selective markers have to be used; for the HfrH strain the choice of *thr*$^+$ *leu*$^+$ was a fortunate one since these two markers turned out to be quite close to the leading end of the injected segment.

What is the reason for the injection by Hfr cells of only a part of their genome? It seems, in fact, that there is no hard and fast limit to the length of the segment that can be injected. In a very small minority of conjugating pairs the whole of the Hfr genome may enter the F^-cell. However, the injection of DNA may be broken off at any time by spontaneous rupture of the connection between the conjugating cells. In the case of the HfrH strain the sequence seldom proceeds much beyond *gal*. In experiments in which conjugation is not interrupted, and in which selection is made for the leading marker, the frequencies in which the unselective markers appear in recombinants decrease in sequence from *azi* to *gal* (Table 3). The same effect is seen in interrupted mating experiments when a long time is allowed for mating (see the 60-minute points, Fig. 15). This *gradient of incorporation*, as it is called, is explained by the decreasing chance of a marker obtaining entry to the F^- cell with increasing distance from the leading end characteristic of the particular Hfr strain. It provides an alternative criterion for the determination of the order of markers on the chromosome.

In order to appear in a stable recombinant it is not sufficient for an Hfr marker merely to enter the F^-cell—it has to become integrated into the chromosome, displacing the homologous DNA of the F^-parent. For this to happen *two* crossovers are required, one on each side of the marker. If selection is made for recombinants which have inherited a relatively late-entering Hfr marker it can be assumed that all preceding non-selected Hfr markers entered the F^- cell and thus had a chance of being included in the recombinants. In so far as they are not so included it must be because crossovers occurred between them and the selected Hfr marker. Again, if the non-selected Hfr markers are close to each other the chances are that they will both be included or both not included in a recombinant, while the further apart they are the more likely it is that one will be included without the other.

Using this sort of reasoning, markers can be ordered on the basis of the frequency of crossing-over occurring between them. Fig. 17 illustrates this method of mapping. Cross-overs between a donor Hfr fragment and the recipient F^- chromosome turn out to be rather

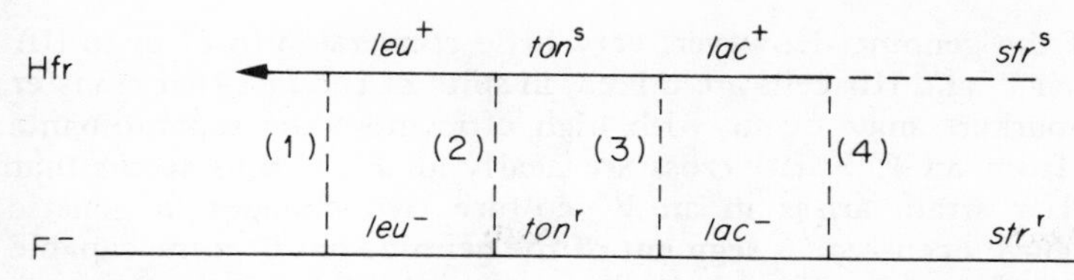

Regions in which exchanges occur	Products	% of total lac^+ str^r recombinants
1, 4	leu^+ ton^s lac^+ str^r	62
2, 4	leu^- ton^s lac^+ str^r	8
3, 4	leu^- ton^r lac^+ str^r	30
1, 2, 3, 4	leu^+ ton^r lac^+ str^r	<1

FIGURE 17 To illustrate formation of different recombinant types by exchanges (cross-overs) in different positions following injection of a DNA segment from Hfr to F^-. In this example the cross was as in Fig. 15 and selection was made for lac^+ from the Hfr donor and for str^r from the F^- recipient; the inheritance of two markers injected *before lac* is shown. The relative distances between genes *leu, ton* and *lac* and between *leu* and the leading end of the Hfr segment are indicated by the relative frequencies of exchange in the different regions. Based on data of Wollman and Jacob.

numerous (an average of about one for each 5 to 10 minutes of the time map). Thus the method is more useful for mapping fairly close markers than for widely spaced ones, all of which tend to show about 50% recombination with each other. The reason for the rapid approach to a limit of 50% is that when the average number of multiple crossovers between markers becomes high the chance of an *even* number of exchanges occurring between the markers tends to become equal to the chance of an *odd* number. In order to achieve net recombination between two markers either one exchange or an odd number of exchanges must occur between them—even numbers, in effect, cancel out.

The inheritance of the fertility factor F demands special consideration. F is remarkable in being capable of infective transmission from cell to cell, like a non-lethal virus. In a mixture of F^+ (non-Hfr) and F^- cells the latter become quite rapidly, and almost completely, converted to the F^+ condition. This transfer of the F factor is accompanied by transfer of other genetic markers in only a very small minority of cases. The F factor behaves, in fact, as a free infective particle independent of the rest of the genome. However, very little conversion to F^+ or to Hfr occurs when F^- and Hfr cells are mixed, in spite of the fact that transfer of other markers may occur with high efficiency; the recombinants recovered from an $F^- \times$ Hfr cross are nearly all F^-. It thus seems that when an Hfr strain arises in an F^+ culture two changes in genetic properties have occurred. A segment of the genome has become capable of being transferred to F^-, and simultaneously the F factor has become non-infective.

One other observation helped to suggest an explanation for these correlated changes. Wollman and Jacob showed that if selection is made for recombinants receiving an Hfr marker close to the extreme trailing end of the genome (the end which is very seldom injected by the usual Hfr strains), then a high proportion turn out to be Hfr rather than F^-. It appeared, then, that a transmissible factor determining the Hfr condition is located at or near the trailing end of the Hfr genome. It was inferred that this factor was the F factor in an integrated state. That F was still present in Hfr strains was indicated by the fact that ordinary F^+ cells arose within them with low frequency. The point of attachment of the F factor, which could be different in each newly arising Hfr strain, was apparently the point at which the circular 'chromosome' breaks prior to injection into an F^- cell. Following breakage, the F factor behaved as if still attached to one side—the trailing side—of the break. All subsequent work has confirmed the essential correctness of this picture, with one qualification. It now seems that the break which initiates chromosome transfer occurs not to one side of the integrated F-factor, but within it, so that a small part of F is injected at the leading end of the donor chromosome while the rest trails at the distal end.

Extraordinary as its properties may seem, F is not unique; as we shall see in chapter 7, there is a whole class of genetic elements called *transmissible plasmids*, of which F is but one example.

What is the mechanism of chromosome transfer from the Hfr to the F^- cell? There is now good evidence to show that it is coupled to DNA synthesis. The break which occurs at the site of the integrated F is believed to be in one strand of the DNA duplex leading to peeling off the broken strand which then passes through the inter-cell bridge (possibly the pilus). According to one hypothesis the transfer is 'driven' by DNA synthesis, either in the Hfr cell, actively displacing the transferred strand, or in the F^- cell where a new strand complementary to the transferred one is synthesized.

TRANSDUCTION – GENETIC TRANSFER THROUGH VIRUS INFECTION

Generalised transduction

In 1952 Zinder and Lederberg, working with *Salmonella typhimurium*, demonstrated that a certain bacterial virus, bacteriophage P22, was able to transmit *bacterial* genetic material from one bacterial strain to another. *Transduction,* as this phage-mediated genetic transfer was called, has profound significance for the understanding of the relationship between viruses and their host cells. and this aspect will be touched upon in chapter 7. In this chapter, I shall merely describe how the process can be made to work, and how it can be used as a tool for mapping the bacterial genome.

The phages which are capable of transducing are all *temperate*—that is to say they are capable of establishing themselves in a host bacterial cell without killing it. A temperate phage may remain in stable association with a bacterial strain for an indefinite time, multiplying in step with the bacterial cells. A bacterium harbouring a temperate phage is called *lysogenic,* because the phage can be induced to multiply rapidly, leading to dissolution or *lysis* of the host cell, by a variety of treatments of which the most convenient is irradiation with ultraviolet light. Bacteria newly infected with a temperate phage may either lyse at once or become lysogenic, the relative probabilities of the two responses depending on the conditions.

To bring about transduction, bacteria of the strain to be used as genetic donor are infected with a suitable phage. After those cells which are going to lyse have done so, the surviving bacteria are killed by the addition of chloroform, and the dead cells and bacterial debris are removed by centrifugation. The phage particles which are left in suspension are used to infect the recipient bacterial strain. After the phage has been absorbed by the bacterial cells the latter are diluted to a suitable concentration and spread on an agar medium which is selective for transduced cells. In the most usual experimental design the recipient strain has an absolute nutritional requirement for some substance which the donor strain can make for itself. Then a medium devoid of the substance in question will only permit growth of cells which have acquired the synthetic ability through the transducing phage.

Only a small fragment of the bacterial genome can be carried in any one phage particle, and the general tendency is for only one genetic marker from the donor strain to be transduced at a time. Where the donor has a second distinguishing marker which is not selected for, this is usually not transferred along with the selected marker. Simultaneous transduction of two markers only occurs when they happen to be closely linked.

As an example of the way in which transduction can be used in genetic mapping we will take an investigation by Yanofsky and Lennox, who used bacteriophage P1 as a means of mapping mutations affecting the synthesis of tryptophan in *E. coli.* The mutant strains included four categories, distinguished by the symbols *trp A*, *trp B*, *trp C* and *trp E*. Tryptophan is normally synthesised in the cell through the reaction of serine with indoleglycerol phosphate, which in turn is derived from anthranilic acid. The abilities of the different mutants to accumulate these intermediates, or to use them for growth, are summarised in Table 4. The data may be interpreted as indicating that each mutant class is unable to carry out one step in the biosynthetic sequence. Each mutant is able to convert to tryptophan only those intermediates coming after the 'block', while the last intermediate before the block tends to be uselessly accumulated.

TABLE 4
Properties of tryptophan-requiring E. coli *strains*

	Growth on minimal medium plus:				
Strain	**No supplement**	**Anthranilic acid**	**Indole**	**Tryptophan**	**Substances accumulated**
wild	+	+	+	+	None
trp E	–	+	+	+	?
trp C	–	–	+	+	Anthranilic acid
trp A	–	–	+	+	Indoleglycerol
trp B	–	–	–	+	Indole

Interpretation in terms of blocks in the biosynthesis:

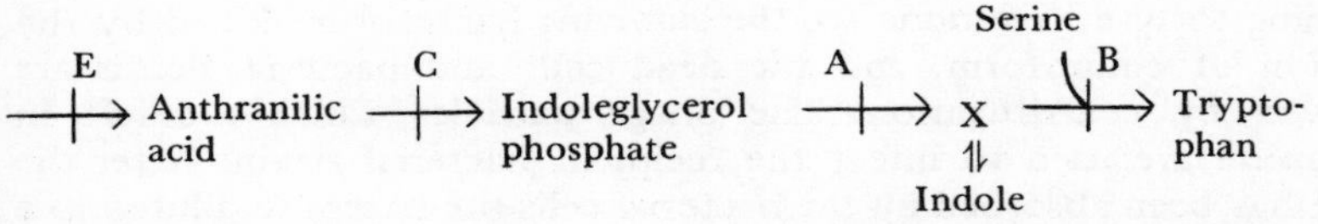

E —|→ Anthranilic acid C —|→ Indoleglycerol phosphate A —|→ X (⇅ Indole) Serine B —|→ Tryptophan

N.B. X is an indole-enzyme complex; free indole is not a normal intermediate in the reaction.

The problem was to determine the linkage relationships of the four kinds of mutant. It was easy to demonstrate that all the mutant sites were rather closely linked. If P1 phage was grown on a *trp E* strain and used to infect *trp B* cells, transductants which were wild with respect to the *trp B* marker could be selected on minimal medium supplemented with anthranilic acid, a substance which will support growth of *trp E* but not of *trp B* mutants. If the *trp B* and *trp E* sites were far apart, the *trp* B^+ transductants would seldom or never have received the *trp E* marker from the donor strain. In fact, however, between 80 and 90% of the *trp* B^+ transduced cells carried *trp E*, and failed to grow in the absence of anthranilic acid. This result can be described by saying that the *trp B* and *trp E* sites show 80–90% co-transduction, and such a high value is indicative of close linkage.

Co-transduction frequencies for other pairs of sites, determined by analogous experiments, were 80–90% for *trp A* and *trp E*, 93–94% for *trp C* and *trp E*, 91–96% for *trp B* and *trp C*, and 97–98% for *trp B* and *trp A*. Thus the probable linkage order was *E–C–(A,B)*, the last two markers being too close together for their order relative to the others to be clear. The order of *trp B* and *trp A* was established by the more critical method of the *three-point test*.

The first step in carrying out such a test was the isolation, by means of a transduction experiment, of a double mutant *trp B trp E* strain. This strain was used as genetic donor with *trp A* as recipient, and transductants with *trp* A^+ from the donor and *trp* B^+ from the recipient strain were selected on minimal medium plus anthranilic acid. Thus an exchange between the *trp B* and *trp A* sites was made obligatory, but, because of the presence of anthranilic acid, there was no selection for or against the *trp E* marker.

In order for the *trp* A^+ marker from the donor to be integrated into the recipient genome, a second exchange is necessary. If the order of sites had been *E–A–B*, this second exchange would have to be to the left of *A*, and would most probably be to the left of *E*, since *A* and *E* show over 90% co-transduction in two-point experiments. Thus the majority transductant class would be $E–A^+–B^+$ due to the following pattern of exchanges:

Donor ——— E ——— A^+ ——— B ———

Recipient ——— E^+ ——— A ——— B^+ ———

On the other hand, if the order is *E–B–A* then the second exchange must be to the right of *A*, and the commonest class of transductant will be $E^+–B^+–A^+$ thus:

Donor ——— E ——— B ——— A^+ ———

Recipient ——— E^+ ——— B^+ ——— A ———

In fact, 90% of the selected transductants were *trp* E^+ rather than *trp E*, and so the order *E–C–B–A* was strongly favoured, and was confirmed by several other three-point tests.

The situation is a little more complicated than has been indicated so far, since each of the four classes of mutant include numerous different sites. All the sites in each class, however, fall within a short segment not overlapping the segments occupied by mutants of other classes, so the order arrived at above is the order of the four functionally distinct segments as well as the order of the four particular sites used in these experiments.

Since the fragments of genetic material carried by a suspension of transducing phage particles are always small compared with a complete bacterial genome, any attempt at mapping the entire genome by transduction experiments would have to be based on piecing together a large number of short overlapping sequences of co-transduced markers. This is hardly a feasible procedure with P1, and in *E. coli* the time-of-entry method described in the preceding section has been relied upon for the overall map, with transduction analysis reserved for determination of the fine detail. In the case of the *Bacillus subtilis* phage BSP1 the situation is rather more favourable; this rather large virus has a head large enough to contain as much as 8% of the bacterial genome and thus can be used to demonstrate linkages between genes which are relatively far apart.

Specialised transduction

Soon after the discovery of generalised transduction mediated by P22, Lederberg and his associates found that another temperate phage *lambda* (λ), which was harboured in latent form by the original K12 strain of *E. coli*, had the special property of transducing the *gal* genes but, so far as could be seen at that time, no other genes of the bacterium. It subsequently turned out that this special transducing ability was due to the actual *integration* of the lambda genome into the bacterial chromosome, at a site close to *gal*, and the subsequent occasional faulty excision of the integrated genome (called a *prophage*) with some adjacent bacterial genes attached to it. In fact a number of genes besides *gal* can be transduced by lambda, the best known being the *bio* cluster, controlling the synthesis of the vitamin biotin, which is also close to the lambda integration site on the other side of it.

The discovery of the chromosomal integration of the λ genome was due to Appleyard and was later extended and confirmed by Jacob and Wollman. In the key experiments Hfr and F^- strains, one of them lysogenic with respect to lambda, were distinguished by a series of markers between *thr* and *str* (cf. Fig. 16). When the F^- parent was lysogenic and selection was made, in the classical manner, for thr^+ leu^+ from the Hfr parent and str^r from the F^- parent, a proportion of the selected recombinants proved to have lost the latent virus. This *absence* of lambda was inherited from the Hfr parent as if closely linked to *gal*. In a reciprocal experiment with the Hfr as the lysogenic parent a quite different result was obtained. At the time after initiation of conjugation at which the Hfr *gal* marker was expected to enter the F^- parent a proportion of F^- cells began to undergo lysis with release of infective lambda particles; thereafter few or no recombinants could be recovered. It seemed that lambda was being introduced into the F^- cells closely linked to *gal* and was immediately multiplying in these cells to cause lysis. This phenomenon, known as *zygotic induction,* is evidently due to the absence of immunity to lambda of bacterial cells into which the virus DNA is newly introduced. We may note in passing that the reason for relative immunity to lysis in a stably lysogenic strain is that, after integration, the phage genome produced a specific *repressor* (analogous to the protein repressors of bacterial gene action to be mentioned in Chapter 6) which inhibits its own excision from the chromosome and free multiplication and development. The possibility of stable integration of lambda in a newly infected cell must depend on the production of repressor sometimes happening just fast enough to prevent the uncontrolled multiplication of the phage. The inactivation of this repressor is the probable reason for the inducing effect of ultraviolet light on lysogenic cells, mentioned above.

The general explanation both for the chromosomal integration of lambda and for its special transducing ability was devised by Campbell and has been confirmed in many ways which cannot be described here. Briefly, the mechanism depends on the ability of the lambda genome to form a closed circle after its injection (in linear form) into the host cell.

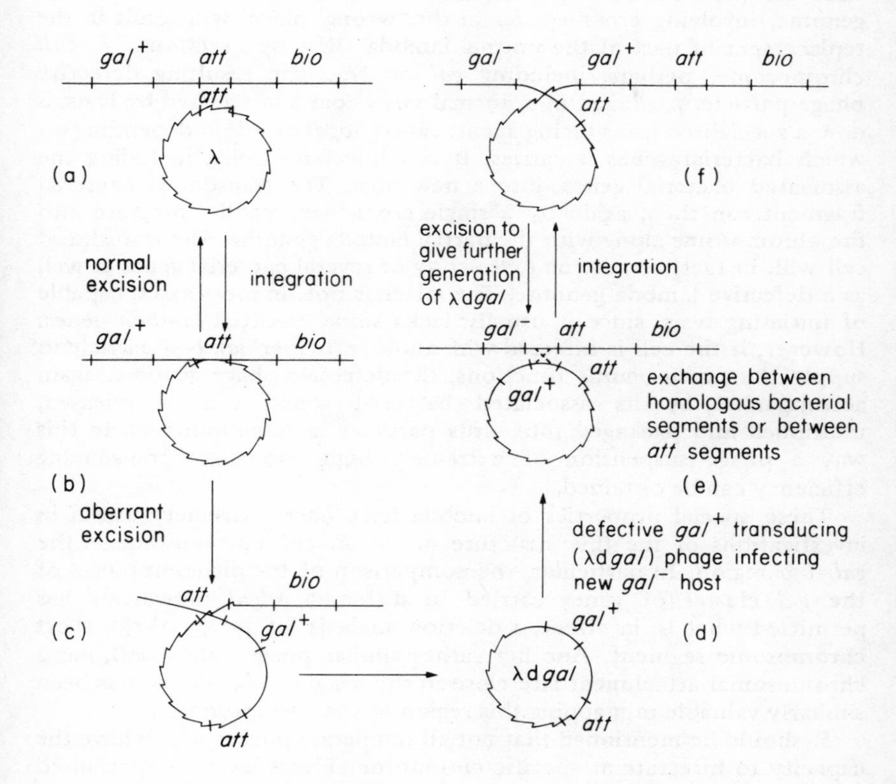

FIGURE 18 Campbell's hypothesis to account for the integration of bacteriophage *lambda* into the *Escherichia coli* chromosome. The *lambda* genome (shown as zig-zag line) forms a closed loop like that of *E.coli* itself (a). Segments of genetic material (shown as *att*) are postulated as specific regions for exchange between *E.coli* and *lambda*. Pairing between these segments in bacterial and phage genomes, with crossing-over, leads to the integration of the *lambda* into the *E.coli* loop (b). The freeing of the prophage at the onset of free multiplication of *lambda* is taken to involve a reversal of this process. Pairing and crossing-over other than between the *att* regions (c) may lead to excision of a defective, *gal*-transducing phage genome λd*gal* (d) which, on infection into another host cell, can readily integrate following pairing either between *att* regions or between a segment common to λd*gal* and the chromosome (e). The result is the introduction of a second *gal* segment into a cell already possessing one (f) and transduction to gal^+ if the recipient cell, as shown here, was originally gal^-.

A single cross-over event (mediated by a specific lambda recombination enzyme) between a special region of lambda DNA and a corresponding attachment site on the bacterial chromosome will, as Fig. 18 shows, result in the linear integration of the virus genome into the chromosome. A reversal of the process, following induction, can release the virus for independent replication. Rare *faulty* excision of the phage genome, involving crossing-over in the 'wrong' place, will result in the replacement of part of the normal lambda DNA by a section of *E. coli* chromosome, perhaps including *gal* or *bio*. The resulting defective phage particle, packaged in a normal virus coat and released by lysis, is now a specialised transducing agent, called λd*gal* or λd*bio* depending on which bacterial genes it carries. It can inject its DNA, including the associated bacterial genes, into a new host. The transduced bacterial fragment can then, again by a single cross-over, readily integrate into the chromosome along with the partial lambda genome. The transduced cell will, in fact, acquire an *extra copy* of several bacterial genes as well as a defective lambda genome. The latter is not, in most cases, capable of initiating lysis, since it usually lacks some essential *lambda* genes. However, if the cell is infected with another 'helper' *lambda* particle to supply the missing virus functions, the defective phage genome, again accompanied by its associated bacterial genes, can be released, multiplied and packaged into virus particles in large numbers. In this way a phage suspension of extremely high specialised transducing efficiency can be obtained.

These special properties of lambda have been extremely useful in investigations of the fine structure of the *E. coli* chromosome in the *gal–bio* region. In particular, the comparison of the different pieces of the *gal* cluster of genes carried in different λd*gal* derivatives has permitted what is, in effect, a deletion analysis (cf. p. 49) of this short chromosome segment. Another rather similar phage called ϕ80, has a chromosomal attachment site close to the *trp* gene cluster and has been similarly valuable in mapping this region of the chromosome.

It should be mentioned that not all temperate phages which have the capacity to integrate at specific chromosomal sites have the specialised transducing property. Phage P22 of *Salmonella*, for example, is now known to have a specific site of integration but does not, for some reason, behave in the same way as *lambda* in forming defective transducing particles. Lambda, for its part, does not have the ability, essential for a generalised transducing phage like P22, of packaging random fragments of purely *bacterial* chromosome in its protein coat during phage assembly prior to lysis.

MAPPING BY TRANSFORMATION WITH DNA

Transformation by preparations of free DNA is quite similar to transduction in the opportunities which it gives for genetic mapping. As in transduction, the pieces of genetic material transferred in transformation experiments are quite small compared with the complete bacterial genome. Consequently, only genetic markers which are closely linked are likely to be carried on a single piece of transforming DNA, and the chance of simultaneous incorporation of two pieces of DNA is extremely low at the DNA concentrations usually used. A number of close linkages have been demonstrated in *Diplococcus pneumoniae* and *Bacillus subtilis* through observations of simultaneous transformation, but a coherent map of the whole genome can hardly be obtained by this method.

In *Bacillus subtilis*, however, transformation has been used in various ingenious ways to determine the sequence in which the genes replicate. Since DNA synthesis always starts at one point on the chromosome and proceeds from that point until all the genes are replicated, the sequence of replication should be the same as the gene sequence determined by any other mapping method. It is possible to bring about synchronous division in *B. subtilis* cultures so that all cells in the culture are replicating a particular gene at roughly the same time. As each gene doubles, transforming DNA prepared from the culture should become doubly effective in transformation with respect to that gene in proportion to others. Careful quantitative assays of transforming activity for a series of genes at different stages of the division cycle can thus provide an overall map of the genome.

Another method which has been successfully used is that of 'density transfer' in which a synchronised culture is grown on medium containing heavy water (D_2O), so that all of its DNA becomes uniformly heavy, and is then transferred to medium containing ordinary H_2O. At different times after the transfer, transforming DNA is prepared, fractionated according to density by the density-gradient method described in the preceding chapter (see p. 16), and the distribution of transforming activities for different genetic markers between the different density fractions is determined. As each gene is replicated it moves from the heavy DNA fraction, with deuterium in both strands, to the 'half-heavy' DNA fraction with one newly synthesised strand lacking deuterium. The sequence in which the genes undergo this density shift gives the gene order.

Reassuringly enough, these novel procedures give the same map as the more conventional transduction analysis.

GENETIC MAPPING IN BACTERIOPHAGES

Some of the most important advances in the analysis of genetic fine structure have depended on mapping the genetic material of bacteriophages. The first phages to be studied from this point of view were T2 and T4, which attack *Escherichia coli.* These two viruses are very similar and need not be discussed separately.

Bacteriophage particles can be identified individually through the circular clearing, or *plaque,* which each one can produce by its multiplication in a lawn of bacterial cells on the surface of a dish of agar medium. The study of bacteriophage genetics is made possible by the existence of mutants which form plaques of unusual size or degree of clarity, or which fail to form them on certain bacterial strains.

Suppose, to take an actual example we infect *E. coli* cells simultaneously with two mutant strains of T2. One of these strains has the mutant trait of *rapid lysis*, giving unusually large plaques and symbolised by r. The second strain has an extension of its host range which permits it to form plaques on *E. coli* strain B/2 which is resistant to wild type T2; this host range character is represented by the symbol h. The r strain is like wild type with respect to host range (h^+), while the h strain forms plaques of normal size on susceptible bacteria (r^+). The phage particles liberated upon lysis of the mixedly infected bacteria are spread on solid medium with a large excess of a mixture of ordinary T2-sensitive and B/2 bacteria. On a lawn of cells of this mixed composition h phage particles form clear plaques since they are able to lyse both types of bacteria, while h^+ particles form turbid plaques because a background of unlysed B/2 cells is left after the sensitive cells have been lysed. The r and r^+ types can readily be distinguished by plaque size. What is actually found is that not only are the expected r^+h and rh^+ plaques formed, but anything up to about 30% of the two recombinant types rh and r^+h^+ also appear (Fig. 19). Virus isolated from

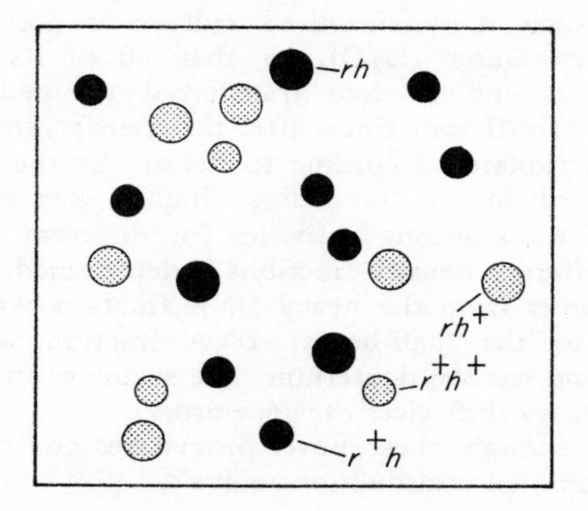

FIGURE 19 The appearance of plaques formed by bacteriophage T2 particles, released from *E.coli* cells mixedly infected with particles of the types $r\ h^+$ (rapid lysis with large plaques, restricted host range) and $r^+\ h$ (normal lysis with smaller plaques, extended host range). The plaques were formed on a lawn of a mixture of cells partly susceptible to both h and h^+ phage, and partly susceptible only to h; the h particles form clear plaques on such a mixture while the h^+ particles form turbid plaques owing to background growth of resistant bacteria (turbidity is indicated here by stippling). As a result of the mixed infection about 30 per cent of recombinant (rh and r^+h^+) particles, as well as 70 per cent of parental-type (r^+h and $r\ h^+$) particles are released from the lysed cells. Redrawn from A. D. Hershey and R. Rotman, *Genetics*, 34, 44-71, 1949.

single plaques of recombinant type subsequently breeds true to type, so there is no doubt that a true genetic recombination is involved.

The theory of recombination in bacteriophages is quite complex, since a mixedly infected bacterial cell is not equivalent to a cross between two individuals on a single occasion, but is more akin to a population cage in which there is opportunity for several rounds of mating between individuals drawn at random from the population. Nevertheless, some of the same principles of mapping apply as in bacteria and higher organisms. If the genetic markers can recombine in mixed infections then they must be at different sites in the genome, and a higher frequency of recombination indicates a greater distance between sites than does a lower frequency.

Recombination percentages in bacteriophage do not represent the probabilities of genetic exchanges at a particular point in time. Levinthal and Visconti, in an experiment in which, by a special technique, they induced mixedly infected bacteria to lyse after different times of infection, showed that recombination frequency increased as lysis was delayed. This implies that exchanges can occur throughout the period of virus multiplication within the bacterium, so that recombinants accumulate with time.

In spite of such complications, when the ratio of the two types of infecting virus particles, the ratio of virus particles to bacterial cells, and the conditions of incubation of the infected cells are all kept constant, the amount of recombination shown by two phage genetic markers is reasonably reproducible from experiment to experiment. When linkage is fairly close, the amount of recombination shown by two markers can be taken as a rough measure of their physical distance apart. Thus by analysing a number of crosses involving different combinations of markers a linear linkage map can be built up.

The T4 genetic map resembles the *E. coli* map in forming a closed loop or circle. The physical basis for this circularity is, however, rather different. The DNA of a T4 particle is a single enormous molecule of 1.3×10^8 daltons, but it is a linear *two-ended* molecule and not a closed loop. The mapping circularity is due to the fact that different T4 particles differ from one another in a very interesting and systematic way. It is as if the stretch of DNA included in each T4 head is a fixed length cut at random from a much longer DNA molecule in which linear copies of the phage genome, each one in the same orientation and with the genes in the same order, are joined end to end. The amount present in a phage particle is about 2% longer than is necessary to include a complete genome, so each particle has the block of genes occurring at one end of its DNA repeated at the other. This situation is called *terminal redundancy*.

What apparently happens when the terminally redundant DNA of a T4 particle replicates within the infected cell is that multiple-length DNA molecules, representing several copies of the T4 genome joined head-to-tail, are formed. Whether these *concatamers*, as they are called, are produced by head-to-tail recombination between the terminally redundant sections of linear molecules after replication, or whether they are generated by continuous peeling-off of a multiple replica from a circular molecule (which could be generated by head-to-tail recombination within a single terminally redundant molecule) is not agreed at the time of writing. The latter mechanism would be reminiscent of the postulated mode of transfer-replication in Hfr *E. coli* cells (cf. p. 38) and similar to the replication of certain other phages such as ϕX174 (cf. p. 21) but there is some evidence against it in the case of T4.

However it is formed, the catameric DNA is evidently cut up by enzyme action into phage-sized pieces. The distance between cuts is fixed, but they can occur anywhere; thus the phage DNA has no constant ends. Taking the whole population of phage molecules the spatial relationship between the genes can only be represented by a circle (Fig. 20).

One of the most impressive exercises in mapping in T4, or in any other genetic system, has been performed by Benzer, using a large number of mutants of the *rII* series.

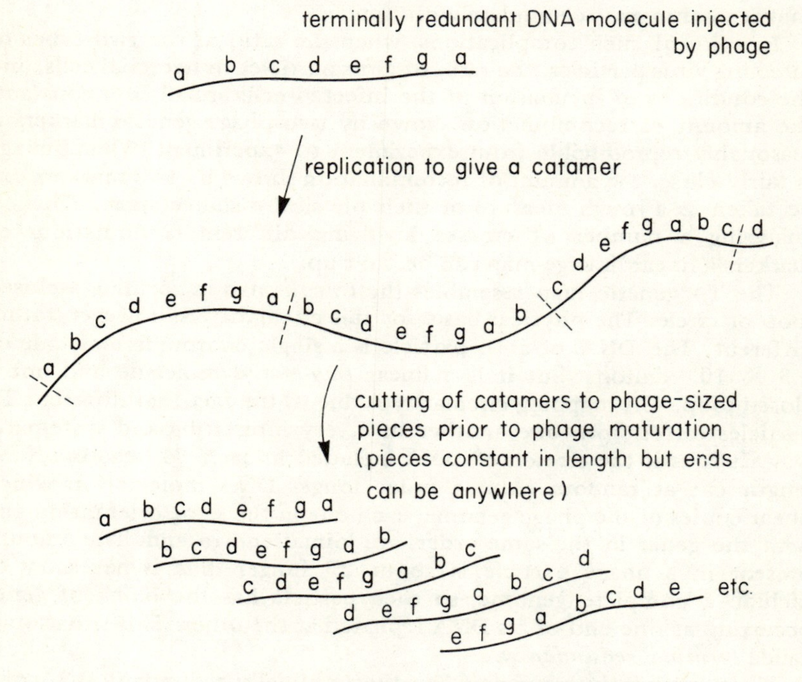

FIGURE 20 The basis of circularity of the bacteriophage T4 genetic map.

There are three quite distinct series of *r* mutants, all forming distinctive large plaques on *E. coli* strain B. The *rII* series are distinguished by the fact that they almost entirely fail to grow on another *E. coli* strain, strain K[1]. Within the *rII* series two distinct classes exist. These classes, called A and B, differ functionally, since any member of one class can *complement* any member of the other. By complementation is meant the ability of a pair of genomes to make good each other's deficiencies, and in the present example is shown by the ability of a *mixture* of virus of types *rII*-A and *rII*-B to multiply in and lyse cells of strain K. Evidently the function, essential for growth in K, which is lacking in A mutants, can be supplied by B mutants in the same cell, and *vice versa.*

Benzer employed two kinds of technique for mapping the *rII* mutants. The first depended on quantitative estimates of recombination frequencies when mutants were 'crossed', by mixed infection of *E. coli* strain B cells, in many different pair-wise combinations. The number of infective particles produced by such mixed infections was assayed by plaque counts on strain B, while the number of r^+ (i.e. wild-type) particles among them was determined by plaque counts on K. Most *rII* mutants, even in single infection, give some low frequency of r^+ progeny, often of the order of $10^{-8}-10^{-6}$. This is attributed to spontaneous reverse mutation. However, the frequency of r^+ progeny arising from mixed infection with a pair of different *rII* mutants is nearly always very much higher—usually in the range 0.02–6%. This enhanced r^+ frequency as a result of mixed infection with two *rII* mutants must be due to the two mutant sites being different, so that each mutant has a wild type version of the genetic site which is mutant in the other.

Very occasionally, a cross between two *rII* mutants gives no r^+ recombinants at all, and this indicates that the two mutant sites involved are identical, or overlap. T4 mutants which are relatively far apart in the genome give as much as 40% recombination, so it is evident that all the *rII* mutant sites are relatively close together. Using the recombination percentages as an indication of distance, the sites can be arranged in an approximate linear order. The results are consistent with all the A mutants being in one genetic segment, and all the B mutants being in another. The two segments appear to be contiguous but non-overlapping. These segments, for reasons which will be explained in Chapter 5, are called *cistrons,* and they correspond to what most geneticists mean by *genes.*

The second technique used by Benzer for mapping the *rII* sites was more novel, and more rigorous. Among the *rII* mutants which he isolated was a large class with two distinguishing properties: (a) unlike most of the other *rII* mutants, they were completely incapable of reverse mutation to wild type, and (b) they each failed to give

[1] The full title of this strain is K12 (λ), and it carries lambda bacteriophage in latent form (cf. p. 42).

recombinants when crossed to any one of a group of other mutants, which occupied a distinct segment of the map and were able to recombine with each other. These two peculiarities pointed to the conclusion that this special class of mutants was due to more or less extensive *deletions* of genetic segments, each one overlapping a number of mutational sites.

When two such deletion mutants are crossed with each other, whether or not they give recombinants will depend on whether or not the deleted segments of the two mutants overlap. By crossing together a set of deletion mutants in many different combinations, Benzer was able to map the deletions in a completely unambiguous linear order. The beauty of this method is that it does not depend on quantitative estimates of recombination frequencies, which may be somewhat variable from one cross to another; all that is necessary is to determine whether each cross produces *any* r^+ recombinants in excess of the (usually vanishingly small) background due to reverse mutation.

The series of overlapping deletions, once established, served to divide the *rII* A and B genes into a number of short segments. Any newly arising mutant can be quickly placed within one or other of these segments on the basis of inability to give recombinants when crossed to deletion mutants overlapping the segment concerned. Fig. 21 illustrates the principle. Benzer worked out a simple, rapid, yet extremely sensitive technique for doing this. A dish of medium is seeded thickly with a mixture of *E. coli* strain B and strain K cells in a ratio of about one to 100, and a large number (about 10^8) of virus particles of one of the standard deletion types. At various well separated points on the surface of the plate, drops of suspensions of various single *rII* mutants are applied. Both mutant types present in each drop are able to multiply enormously at the expense of the minority of B cells; the subsequent fate of the K cells depends on the genetic relationship of the two mutants. Three kinds of result are observed in different cases. Sometimes all the bacteria in the area of the drop lyse to give a large circular clearing. This indicates that the two mutants fall within different genes or cistrons, so that both types can multiply freely by complementation in mixedly infected K cells. A second type of result is a complete absence of any plaques in the heavy background of K cells; this indicates that the two mutants not only fall within the same gene but that they actually overlap, making recombination impossible. Thirdly, a number of small plaques may develop within the area of the drop, indicating that r^+ viruses are being formed as a result of the mixed infection of the B cells, and thus that the mutant under test is at a site which is *not* overlapped by the standard deletion. The test is capable of detecting as few as 0.01% r^+ recombinants. Once the new mutants have been allocated to segments of the map, their order within the segments can be established by quantitative estimates of recombination frequencies. The extraordinary amount of genetic detail which can be resolved by this system will become evident from the discussion in the next chapter (cf. Fig. 24).

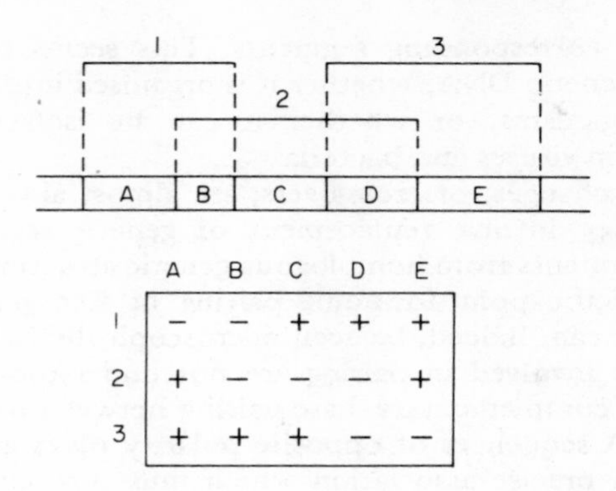

FIGURE 21 To illustrate the principle of allocating bacteriophage mutants to genetic segments on the basis of their ability or inability to yield wild type recombinants when crossed to a standard series of overlapping deletion-type mutants. Here deletion mutants 1, 2 and 3 are used to define five segments A–E. The table (below) shows the results of crossing mutants falling within each of the five segments with each of the three deletion mutants; + indicates formation of recombinants and–indicates no recombinants. The map (above) shows the order of the segments so defined.

One use to which the fine structure mapping of the *rII* region has been put is the estimation, in chemical terms, of the size of the smallest amount of genetic material which can be resolved by recombination studies. The total molecular weight of all the DNA in a T4 particle is about 2.6×10^8, corresponding to 3×10^5 nucleotide pairs. The total length of the T4 genetic map, in terms of units of 1% recombination, is of the order of 1000, while the smallest recombination frequencies recorded by Benzer were about 0.02%; in fact, still lower frequencies, at the very edge of detectability above the background due to reverse mutation, are now known to be shown by the very closest-linked sites. The value of 0.02% corresponds to about 2×10^{-5} of the total map which, on the simplest possible assumptions, corresponds to 4 nucleotide pairs. While this calculation is extremely rough, it does suggest that the genome of T4 is subdivisible by recombination right down to its ultimate chemical units and that recombination can most likely occur between any two adjacent base pairs.

THE UNIVERSALITY OF GENETIC RECOMBINATION AND ITS MECHANISM

The basis of all genetic mapping lies in the ability of homologous chromosomes, or homologous pieces of genetic material, to undergo exchanges of corresponding segments. This seems to be a universal property of genetic DNA, whether it is organised into chromosomes, as in higher organisms, or whether it can be isolated as free DNA molecules, as in viruses and bacteria.

Genetic exchanges, or cross-overs, are almost always precise, in the sense that they involve replacement of genetic segments by *exactly equivalent* segments from homologous genetic structures. This precision implies a specific point-for-point pairing of like genetic segments—a pairing which can, indeed, be seen microscopically in higher organisms.

The forces involved in pairing are not understood. It is natural to suppose that complementary base-pairing between homologous single-stranded DNA sequences of opposite polarity plays an essential part in the close and precise association which must precede recombination, since we know of no alternative. The difficulty with this idea is that the forces involved in hydrogen-bond formation are very short-range ones; they could stabilise any close homologous alignment brought about by chance during random molecular movements, or by some unknown mechanism, but could not be responsible themselves for such alignment. In eukaryotes the pairing at pachytene depends on the formation of a ribbon-like predominantly protein structure called the *synaptonemal complex,* between each homologous chromosome pair. This complex seems in some quite unknown way to be able to bring about, or at least to stabilize, an exact alignment of chromosomes even though these still seem to be much too far apart to be interacting by hydrogen bonds. Whether anything comparable occurs in prokaryotes is not known, but there is no evidence for it.

So far as the crossing-over which follows close association is concerned, there have in the past been two main types of explanation for it. One is the *copy-choice* hypothesis, according to which DNA replication occurs while homologous DNA molecules are so closely associated that a newly growing strand can switch from one as template to the other, so that it is copied partly from one genome and partly from the other. This hypothesis in its pure form has been ruled out by the demonstration in a variety of eukaryotic and prokaryotic systems that recombination can unite *previously synthesised* DNA molecules of different origins. For example, *lambda* bacteriophage recombinants can be formed virtually without new DNA synthesis when *E. coli* cells are infected with numerous *lambda* particles of two different genotypes; and, in eukaryotic meiosis, the major replication of the chromosomal DNA is completed some time before chromosome pairing and no more major DNA synthesis occurs until after meiosis and recombinant formation is completed. Thus current thinking favours the second type of hypothesis, briefly referred to as the *breakage-rejoining* model.

It is not possible in the space of this book to discuss the several detailed proposals which have been advanced as to how breakage and exchange of homologous DNA segments may occur. Suffice it to say that all of them involve the breakage of single strands of the initially double-stranded DNA molecules, followed by the association in hybrid duplexes of complementary single-strands from the different molecules. Following this formation of hybrid DNA, the different variants of the hypothesis propose different ways in which the initially single-stranded cross-over could lead, through further breakage and a certain amount of local DNA degradation and resynthesis, to the formation of a complete crossover of both strands of the participating molecules. It is fairly certain that some of the enzyme-catalysed reactions necessary for crossing over, involving single-strand breakage, limited digestion, re-synthesis and rejoining, are also involved in the repair of DNA damaged, for example, by ultraviolet irradiation (see next chapter). The main reason for this statement is that several classes of mutant bacteria, abnormally sensitive to ultraviolet light because of defects in DNA repair mechanisms, are also defective in their ability to produce genetic recombinants. The identification of the specific enzymes missing or defective in UV-sensitive, recombination-deficient mutants is currently providing new clues to the solution of the recombination problem.

Finally, it is worth pointing out that the dichotomy, much emphasised in the past, between copy-choice mechanisms involving new DNA synthesis and breakage-rejoining mechanisms not doing so, now seems to be a false one. The overall effect is of breakage-rejoining of presynthesised segments, but in detail, in the immediate vicinity of a crossover, a limited amount of trimming of DNA strands and compensating repair synthesis almost certainly occurs.

One must always keep in mind the possibility that the mechanism of recombination may differ in detail or even in major features as between eukaryotes and prokaryotes, or between bacteria and viruses. Irrespective of the mechanisms involved, however, the fact remains that there is a strong formal similarity between the mapping procedures in all these diverse systems. The basic postulate is that the number of exchange points is limited, so that a map order which permits the most common recombinant classes to be explained as due to single exchanges is more probable than one which would require multiple exchanges to be more frequent.

A COMPARISON OF PROKARYOTE AND EUKARYOTE CHROMOSOMES

There are large differences between viruses, bacteria and higher organisms in the amount of DNA per genome and, very likely, in the complexity of its organisation. In T4 bacteriophage the amount of DNA per particle is equivalent to a single piece of molecular weight 1.2×10^8, while the corresponding figure for *E. coli* is of the orderof 10^9 per nuclear body, and for *Neurospora crassa* about 3×10^{10} per nucleus. In flowering plants and animals the figure is one or two orders of magnitude higher again. In bacteriophage it is clear that the structure represented by the linkage map is a single large molecule of DNA. The same seems to be true of *E. coli*; the autoradiographic evidence of Cairns (reference on p. 139) and several later workers shows that DNA can be freed from *E. coli* cells in large continuous closed loops each equivalent to one nuclear body.

Turning to fungi we find a different situation. Here the genetic map consists of a number of linear linkage groups corresponding to separate linear chromosomes—seven in *Neurospora crassa* and upwards of sixteen in yeast. The amount of DNA in one fungal chromosome is similar to or even less than that in the single chromosome of *E. coli*. In yeast it has been shown that each chromosome contains just one continuous linear DNA molecule; in higher eukaryotes, where the quantity of DNA per chromosome is relatively enormous, the situation is not so clear but is probably essentially similar.

The constant association of the DNA of higher plants and animals with basic proteins of the *histone* class (probably replaced in fungi by other basic proteins of less well-characterised kinds) is presumably connected with the cycle of extreme condensation and elongation undergone by the eukaryote chromosome during mitosis, a cycle which has no known counterpart in prokaryotes. Chromosomal protein in eukaryotes is probably also of significance in relation to the regulation of gene activity, the mechanisms of which are as yet poorly understood in eukaryotes in comparison with the detailed (albeit still incomplete) picture which is beginning to emerge for prokaryotes and which is sketched in chapter 6.

4 Mutation

The maintenance of the very precise organisation of living organisms depends on the high degree of accuracy with which genetic information is transmitted from one cell generation to the next. As we saw in Chapter 2, this accurate transmission finds an explanation in the self-replication of DNA, which is the most general kind of genetic material. However, notwithstanding the impressive stability which species show with regard to their basic systems of organisation, a considerable amount of heritable variation can occur. It is this variation which provides us with the genetic markers which are essential for genetic mapping, and which gives organisms their potentiality for evolutionary change.

What is the origin of the genetic differences which are found between organisms which are obviously of common stock? It is difficult to believe that species sprung into existence already equipped with all the variation which we now find in them, and it is, in fact, easy to show experimentally that new variants are constantly arising at a low frequency. The origin of a new genetic trait is called a *mutation*, and an organism showing the effects of a mutation is called a *mutant*.

SPONTANEOUS MUTATION

It was long a matter for controversy whether mutation tends in itself to have adaptive value—that is, whether those mutations tend to arise which are to the organism's advantage in the prevailing environment—or whether, on the contrary, mutations occur at random with respect to the environment and merely furnish material for subsequent selection. The second view is now generally accepted. As we shall see later in this chapter, many physical and chemical agencies are known which will increase mutation frequency, but in no case do induced mutations seem to be adaptive in relation to the inducing treatment. Furthermore, mutations of all kinds tend to occur all the time in an apparently spontaneous fashion and with no apparent relationship to the demands of the environment.

A classical demonstration of one kind of spontaneous mutation in a bacterial population was made by Luria and Delbrück in 1946. They showed that if a large number of cells of *Escherichia coli* were infected with an excess of bacteriophage T1 most of the cells underwent lysis, but a very small proportion (of the order of 1 in 10^8) were able to multiply and form colonies on plates of solid medium. When these colonies were sub-cultured they were found to consist of cells which were resistant to T1 and were able to transmit their resistance to their descendants. Thus a small number of the originally infected cells were already, or became, resistant to phage through a genetically stable mutation. Now there were two possible views of these mutations. One could suppose that they were caused, though with very low efficiency, by the phage infection. Alternatively, it was possible that the mutants were already present and that the phage merely selected them out. By an ingenious experiment Luria and Delbrück proved that the second alternative was correct.

The experiment consisted of growing, in liquid medium, a series of similar small cultures of bacteria for a fixed time from as nearly as possible identical inocula, and then infecting each culture with an excess of T1 particles and spreading it on a plate of agar medium. The number of phage-resistant cells in the infected culture. For comparison, a similar number of plates were spread with similar numbers of infected bacteria, but this time drawn as samples from a single large liquid culture. If the mutations to phage resistance occurred as a consequence of phage infection then all cells on all the plates should have had an equal chance of mutation. Thus the variation in the numbers of resistant colonies should, in this case, be no greater in the series of plates inoculated with the series of parallel cultures than in the series inoculated with samples from the same culture. If, on the other hand, the resistant colonies were descended from mutations which occurred at random at various times before infection, then the number of mutants in each of the series of parallel cultures would depend on the stage in the growth of the culture at which the mutation or mutations occurred. Those cultures in which, by chance, an early mutation occurred would come to contain a relatively large number of resistant cells at the time of infection, while others, which did not happen to have a mutation at all, would contain none. In short, the variation among different, apparently identical, cultures would be much greater than among different samples from the same culture. This kind of experiment is known as a *fluctuation test*. Luria and Delbrück's results spoke decisively in favour of the hypothesis of random and spontaneous mutations. Table 5 shows some of the data.

Later, in 1952, Lederberg produced a rather more direct demonstration of the spontaneous nature of mutations to T1 resistance. His experiment depended on his technique of *replica plating,* in which a sterile velvet surface, supported on a circular block, is used to transfer cells from an ensemble of colonies on one plate of agar medium to a fresh plate where, after incubation, an identical pattern of colonies is formed provided that the medium will support their growth. The starting point of Lederberg's experiment was a plate of nutrient agar medium heavily seeded with bacteria to give a continuous lawn of cells.

TABLE 5
The fluctuation test of Luria and Delbrück

	Numbers of phage-resistant mutants present in:						
	Samples from the same culture			Different parallel cultures			
Expt. no.	**1**	**2**	**3**	**4**	**5**	**6**	**7**
	14	46	4	10	30	6	1
	15	56	2	18	10	5	0
	13	52	2	125	40	10	7
	21	48	1	10	45	8	0
	15	65	5	14	183	24	303
	14	44	2	27	12	13	0
	26	49	4	3	173	165	0
	16	51	2	17	23	15	3
	20	56	4	17	57	6	48
	13	47	7		51	10	1
							4
Mean	16.7	51.4	3.3	26.8	23.8	26.2	30
Variance	15	27	3.8	1217	84	2178	6620

NOTE: The variance, which is the sum of the squares of the deviations from the mean divided by one less than the number of samples, should be approximately equal to the mean if the variation is due only to chance differences between samples drawn from a single population. From Luria and Delbrück (1943).

A sample of cells from this master plate was transferred with the replicator to a second plate which had been coated with T1 phage particles. A few mutant phage-resistant colonies grew on the second plate. If these colonies stemmed from colonies of mutant cells already present on the original plate, then it was expected to be possible to obtain a culture with an increased proportion of such cells by taking inocula from points on the master plate corresponding in position to the resistant colonies on the phage-coated plate. Lederberg was, in fact, able to do this. A second master plate, seeded less thickly than the first, was prepared from the enriched culture, and replication on to phage-coated medium was repeated. After this cycle of operations had been repeated 3 or 4 times a large proportion of the cells in the enriched culture were T1-resistant, and the master plate could be seeded thinly enough to permit the identification of single resistant colonies on it, following replication. These colonies could be used to establish true-breeding phage-resistant strains. It must be emphasised that these strains descended only from the bacteria on the series of master plates which were *never in contact with phage*. The whole

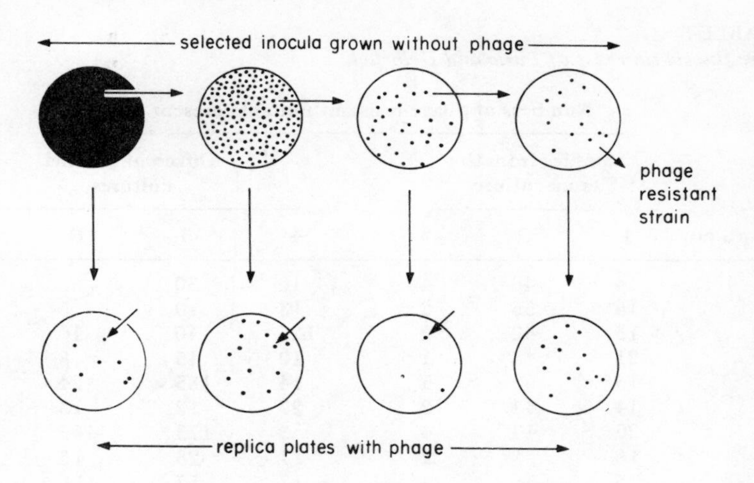

FIGURE 22 Lederberg's method for isolating a strain of *Escherichia coli* resistant to bacteriophage T1, using indirect selection with replica plating. The small diagonal arrows show the T1 resistant colonies by reference to which resistant cells on the phage-free plates were located. At each stage cells were taken from the selected point on the phage-free plate, diluted, and spread on the next plate of the series. The number of steps required in practice would probably be somewhat greater than shown in the figure.

selection procedure was indirect (Fig. 22). There was thus no question of the mutation to resistance having been induced by the virus. Similar demonstrations of the spontaneous origins of bacterial strains resistant to various antibiotics and other drugs were afterwards made.

It may seem extraordinary that bacterial populations should constantly give rise to mutants able to cope with environments which the strain is not currently experiencing, and which it may never have experienced before in its evolutionary history. Nevertheless such mutational versatility does seem to be a fact. It must be remembered, however, that any given kind of mutation occurs extremely rarely, and only gets a chance to be selected because of the enormous numbers of cells present in bacterial populations.

SELECTION OF AUXOTROPHIC MUTANTS

Mutants which will grow under conditions in which the wild type organism will not are easy to select. A good many of the most useful mutants for genetic and biochemical studies are, however, more limited than wild type in their growth abilities. The most widely useful kind of mutant of all is the *auxotroph*, several examples of which have been referred to in earlier chapters. Auxotrophic mutants are unable to grow on the *minimal medium* (consisting, in the case of *Escherichia coli*, simply of a solution of inorganic salts and glucose, solidified if necessary with agar) which suffices for the growth of the wild type. They each need some additional substance, or occasionally a mixture of two or more substances, before growth can occur. The growth requirement may be for a vitamin, an amino acid, or a nucleic acid base or nucleoside. Auxotrophic mutants evidently have lost the capacity to synthesise for themselves the substances which they are found to require.

A very neat method for isolating rare auxotrophs from large populations of *E. coli* was devised by B. D. Davis in 1949. He made use of the fact that penicillin only kills growing bacteria. Thus when a suspension of cells is made in liquid minimal medium containing penicillin, the non-auxotrophic (*prototrophic*) organisms nearly all grow and are killed. The occasional auxotrophic cells, on the other hand, are unable to grow and tend to survive. After an adequate time the surviving cells are centrifuged from the medium, washed free of penicillin, and spread on plates of solidified minimal medium supplemented with whatever substance or group of substances one happens to be interested in. In principle, only auxotrophs responding to the supplement will grow. Though some surviving prototrophs usually appear on the plates, colonies of the desired auxotrophic types usually also occur. These latter can be distinguished by their failure to grow on minimal medium following replica plating. The penicillin selection method has been used for the isolation of hundreds of different kinds of auxotrophic mutant in *E. coli*.

CONDITIONAL MUTANTS

Auxotrophs are themselves *conditional mutants* since they only appear abnormal in certain circumstances—their failure to grow is conditional on the absence of a particular nutrient. They have the limitation that their possible functional defects are restricted to those which can be repaired by substances of low molecular weight which can be absorbed from the growth medium. Other kinds of conditional mutants are sometimes even more informative, since they are not subject to this restriction. Such is the case with *temperature-conditional* mutants. These mutants fail to grow normally only in a certain temperature range—usually at higher temperatures. As we shall see in the following chapter, temperature-conditional (or *temperature-sensitive*) mutants are usually vulnerable to higher temperatures solely because of defects causing thermolability in specific proteins. Virtually any protein, no matter what its function, is likely to be capable of acquiring this kind of defect by mutation. Temperature-sensitive (*ts*) mutants can be identified in bacteria by replica-plating an ensemble of colonies and incubating the replica-plate at close to the maximum temperature permitting growth of the wild type (about 42°C for *Escherichia coli*). Failure of a replica colony to grow under these conditions indicates that it is probably a *ts* mutant.

A second type of conditional mutant is exemplified by the class which was identified first in T4 bacteriophage and later in *E. coli*, and called, for an unimportant reason, *amber*. Amber mutants of T4 were defined as those which grew perfectly well in one strain of *E. coli* (the permissive host) but not at all in another (the restrictive host) It was shown that the permissive host was so by virtue of carrying a gene, a so-called *amber-suppressor*, which nullified the effect of the amber mutation. It was subsequently shown that amber mutations could occur in the *E. coli* genome as well, but likewise only showed their effect in cells lacking an amber suppressor. Other categories of suppressible mutants, notably a second class called *ochre*, were later identified; some suppressor genes in *E. coli* would suppress ambers and ochres while others were specific for ambers. Amber and ochre mutants are of special significance in relation to the punctuation of the genetic code, whereby the information in what are sometimes very long RNA molecules (messengers—see p. 83) are used to specify the structures of polypeptide chains of limited length (see the next chapter).

REVERSE MUTATION

Many mutants, perhaps a majority, are able to revert to the wild type condition by *reverse mutation.* Some kinds of reverse mutations provide the most convenient experimental systems for the study of mutation frequencies. Reversions of auxotrophs to the prototrophic condition, in particular, are very easily selected and counted simply by plating large known numbers of cells of the original auxotrophic strain on minimal medium. In bacteriophage T4 *rII*$^+$ revertants in *rII* strains can likewise be easily estimated through their ability to grow in *E. coli* strain K (cf. p. 49).

It should be pointed out, in this connection, that the distinction between a 'forward' and a reverse mutation can be somewhat arbitrary. Where a mutant has been experimentally induced, and has what is obviously a grossly abnormal or deficient phenotype, it seems reasonable to refer to the restoration of the more normal phenotype (the 'wild type') as reverse mutation. There is, however, no such thing as one invariant and recognisable wild type in any microbial species; all strains actually isolated from the wild (i.e. the world outside the laboratory) doubtless carry genes of more or less recent mutational origin. Furthermore, what appears to be a reverse mutation so far as the phenotype is concerned is not *necessarily* due to a precise reversal of the original change in the genetic material. Modern fine structure mapping, both in T4 bacteriophage and in *E. coli*, has shown that the effect of one mutation may sometimes be partially or entirely overcome by a second *suppressor* mutation, which may, in different cases, either be at a second site within the same gene or elsewhere in the genome.

MUTAGENS AND THE MECHANISMS OF MUTATION

The spontaneous mutation frequency at any one genetic site (DNA nucleotide pair) is usually extremely low, often in the range 10^{-8} to 10^{-10} for each replication of the genome. Greatly increased mutation frequencies can be obtained through a variety of mutagenic treatments.

The longest known and most widely used mutagenic treatments are irradiations of various kinds. The mutagenic effects of X-rays, and other ionising radiations such as gamma rays or fast neutrons, have been known since the mid-1930's. Because of their penetrating power, ionising radiations are mutagenic for all kinds of organisms including higher animals. Ultra-violet light is also mutagenic, but is only effective on bacteria or fungal spores or other free cells whose genetic material is not shielded by any great thickness of ultra-violet absorbing material. No one precise mechanism can be proposed to account for the effects of X-rays, but it is not surprising that they do have a mutagenic effect since they can cause the rupture of many different kinds of chemical linkage and thus, no doubt, damage the genetic material in a variety of ways.

Rather more can be said about the action of ultraviolet light (UV). Its major effect is to cause the formation in DNA of *dimers* by cross-linking between adjacent pyrimidine, especially thymine, residues thus:

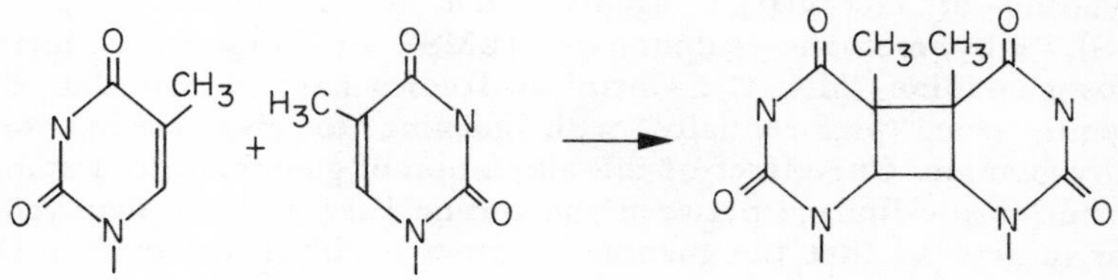

It seems certain that the formation of thymine dimers is a major cause of killing of cells by UV through their disruption of the normal process of replication. Bacterial cells of many kinds, and also yeast, have been shown to possess an efficient *photoreactivating* mechanism for repairing the damage, unlinking the dimers; this repair process is promoted by visible light and is catalysed by a special enzyme. Not only will photoreactivation restore full viability to UV-damaged cells (provided the cells are exposed to the visible light soon enough) but it will also nullify the mutagenic effect of UV. The implication is that the mutation, like the killing, stems from the dimer formation. There are strong indications that the later steps in the sequence of events leading to mutation involve another repair mechanism which comes into play if photoreactivation fails. This second-line defence operates through the enzymic removal of single-strand sequences of DNA nucleotides including pyrimidine dimers and their replacement by new synthesis using the second strand of the DNA duplex as template. This process apparently is prone to 'mistakes', occasionally inserting 'wrong' nucleotides during the repair synthesis. The main evidence for this is that certain UV-sensitive mutants which cannot repair their DNA by this 'cut and patch' mechanism are not susceptible to UV-induced mutation.

The most stimulating results on mutation in recent years have come from studies on the mutagenic effects of chemicals of various kinds. The modes of action of several chemical mutagens have been more or less worked out, and the different degrees of susceptibility shown by specific genetic sites to different mutagens often provides evidence as to the chemical nature of the sites. Many of the pioneering experiments in this field were carried out by Benzer and by Freese on the bacteriophage T4 *rII* series of mutants; the conclusions drawn have, by and large, proved applicable to bacteria and even fungi.

There are two main categories of mutagenic chemicals. The first consists of compounds which can react chemically with isolated DNA and which can induce mutations in free bacteriophage particles or in bacterial transforming preparations. The second consists of *base analogues* which are sufficiently similar to normal DNA bases to substitute for them during DNA replication in a living system.

Some of the most effective mutagens in the first class are *alkylating agents,* of which mustard gas and other 'mustard' compounds were the first to be investigated. Most of those in extensive current use are ethylating or methylating agents, such as ethylethane sulphonate (EES), ethyl-methane sulphonate (EMS) or N-methyl-N′-nitro-N-nitrosoguanidine (MNNG). Chemical studies have shown that these reagents react preferentially with guanine to give 7-ethyl or 7-methylguanine. One effect of this alkylation of guanine is to destabilise the nucleoside linkage between the purine base and the deoxyribose sugar moiety so that the guanine becomes readily lost from the DNA chain. It has been suggested that this is at least part of the reason for the mutagenic effect, since a DNA strand missing a base will, if it replicates at all, presumably be free to bring in any of the four possible bases opposite the gap when the complementary strand is synthesized. It seems more likely, however, that loss of guanine is usually lethal and that 7-alkylguanine brings about mutations without being split from the chain through changes in its base-pairing properties, rather in the manner of the base analogues discussed below.

A reagent which acts in quite a different way is *hydroxylamine.* Freese showed that this chemical reacts with cytosine (or, in T4 DNA, with hydroxymethylcytosine), and he was not able to detect any reaction with any of the other DNA bases. The likely structure of the reaction product is shown in Table 6. If hydroxylamine does indeed act on cytosine in this way the product would probably no longer be able to form a hydrogen-bonded pair with guanine, and might pair instead with adenine at the next replication.

Nitrous acid is well known in organic chemistry for its reaction with amino groups to give hydroxyl groups with the liberation of nitrogen gas. Through this reaction adenine is converted to hypoxanthine, guanine to xanthine and cytosine to uracil. Of these changes the last should certainly be mutagenic, since uracil has the hydrogen-bonding properties of thymine, while the first is probably mutagenic, hypoxanthine seeming more likely to pair with cytosine than with thymine (see Table 6). Like the other reactive chemicals just mentioned, nitrous acid is a potent mutagen when used on free bacteriophage particles, isolated infective RNA from tobacco mosaic virus, or on intact bacterial cells.

The base analogues, on the other hand, do not cause mutations in isolated genetic nucleic acid but only in replicating DNA in living systems. Two base analogues have been used more than any others. 5-Bromouracil (5-BU) is a close analogue of thymine, having a bromine atom in place of the rather similar-sized methyl group. 2-Aminopurine (2-AP) is closely related to adenine, differing from it in having the amino group in the 2- rather than in the 5-position. Both these analogues induce mutations in dividing bacterial cells or in multiplying bacteriophage. It is easy to show that in a bacterium unable to synthesise thymine, either because of the presence of the drug aminopterin, which inhibits the synthesis of methyl groups, or because of the genetic loss of one of the necessary enzymes, 5-BU supplied to the organism is massively incorporated into DNA. The analogue fits almost perfectly into the DNA structure, and cells containing a high proportion of 5-BU are viable to a surprising extent. Such cells tend to show an increased mutation rate, and it has been plausibly suggested that the reason for this is that 5-BU is less specific than thymine in its pairing with other bases. While it would seem to pair normally with adenine most of the time, the increased electronegativity of the bromine as compared with the methyl group could make 5-BU more prone than thymine to undergo the tautomeric change which would permit it to pair with guanine (Table 6). If 5-BU does become incorporated into DNA opposite guanine, it is likely that at the next replication it will behave in the more usual way and cause the incorporation of adenine. Thus an AT pair will become substituted for a GC pair, the entire sequence of events being G-C → G-BU → A-BU → A-T.

2-AP is not incorporated into DNA nearly as readily as is 5-BU, though there is evidence that it is incorporated to a small extent under appropriate conditions. Yet its mutagenicity is of the same order as that of 5-BU. It is likely that 2-AP makes up for its lower probability of incorporation by the greater ambiguity of its behaviour when it is incorporated. If incorporated in place of adenine it may have quite a high probability of pairing with cytosine during replication instead of with thymine. This would bring about an AT to GC base pair change, the sequence of events being A-T → 2AP-T → 2AP-C → G-C.

Table 6 summarizes possible modes of action of the various mutagens just discussed. The ability of the normal bases and the base analogues to form hydrogen bonds are indicated by arrows pointing away from hydrogen 'donor' groups and towards hydrogen 'acceptors'. Only where some of the donor and acceptor groups 'fit' can a hydrogen-bonded pair be formed.

TABLE 6
Possible mechanisms of action of some chemical mutagens

Normal DNA base pairs:

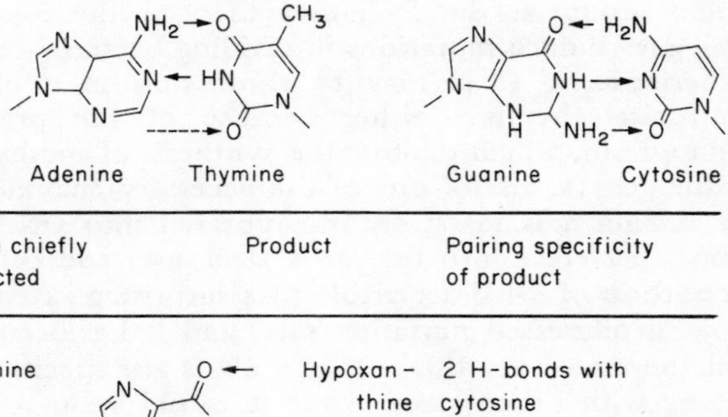

Adenine Thymine Guanine Cytosine

Mutagen	Base chiefly affected	Product	Pairing specificity of product
Nitrous acid	Adenine	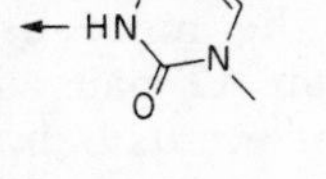 Hypoxan-thine	2 H-bonds with cytosine
	Cytosine	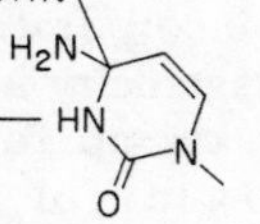 Uracil	2 H-bonds with adenine
Hydroxylamine	Cytosine		1 H-bond with adenine ?

Alkylating agents	Guanine	7-alkyl guanine tending to split from the DNA chain	
2-amino purine (2AP)	Adenine	2-AP substituting for adenine	2 H-bonds with thymine or 1 with cytosine
5-bromo-uracil (5-BU)	Thymine	5-BU substituting for thymine	2 H-bonds with adenine or 3 with guanine

The hydrogen-bonding properties of the various chemical groups and atoms are shown by arrows, which point away from hydrogen donor groups and towards hydrogen acceptor groups.

On the basis of the theory of mutagenesis just outlined one would predict that base analogues would cause mutation in two distinct ways. One way may be called *mutation by mis-incorporation*; for example 5-bromouracil might be incorporated as if it were cytosine, and then proceed to replicate as if it were thymine. Alternatively, if 2-aminopurine, for example, were 'correctly' incorporated as an adenine analogue, and then replicated at some subsequent division as if it were guanine, this would lead to *mutation by mis-replication.* It seems, in fact, that both analogues can act in both ways, since each can usually cause reversion of mutations induced by itself. However, there is a tendency for 5-BU induced mutants to be reverted with much higher frequency by 2-AP than by 5-BU. This suggests that 5-BU induces preferentially one of the two possible base pair transitions. A strong indication of which one this is is provided by Benzer and Champe's observation that the group of T4 *rII* mutants which are induced to revert with high frequency by 5-BU are almost the same as the group which are revertible by hydroxylamine. Since there is evidence that hydroxylamine acts specifically on cytosine or (in T4) hydroxymethyl-cytosine, 5-BU is thought to induce preferentially GC to AT transitions, and thus to act mainly by mis-incorporation. 2-AP seems to be less specific and to act with more or less equal efficiency by either mechanism.

There should be a pronounced difference between mis-incorporation mutation and mis-replication in respect of the time during which the induced mutants appear. Following mis-incorporation of an analogue the mutation is essentially established at once. With mis-replication, however, where the analogue is 'normally' incorporated, mutants will continue to arise, with a constant probability per replication, for an indefinite number of cell generations. Thus if, for example, an auxotrophic mutant of *E. coli* is made to incorporate an analogue during one cell division cycle, and then samples of the culture are plated on minimal medium at various times thereafter to determine the proportion of prototrophic reverse mutants, one may expect either of two results. If reverse mutation is by mis-incorporation there will be a once-and-for-all burst of mutation in the generation following incorporation, after which there will be no further increase in the proportion of revertant cells. If, on the other hand, reversion occurs by mis-replication, the proportion of revertants will continue to increase for an indefinite time as the culture grows.

Fig. 23 illustrates the argument which, as first shown by Strelzoff, provides a way of distinguishing between the two mechanisms of mutagenesis.

All the evidence is consistent with the hypothesis that base analogues can only cause base-pair *transitions* in the restricted sense proposed by Freese; that is, the substitution of a purine for a purine

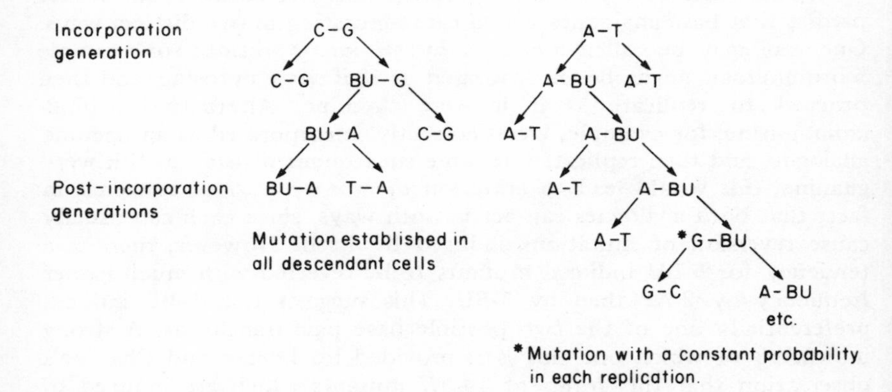

FIGURE 23 The different consequences of two mechanisms of mutation induced by 5-bromouracil.

and a pyrimidine for a pyrimidine. In fact revertibility with a base analogue is now widely regarded as diagnostic for a transition mutation. Among the *rII* mutants of bacteriophage T4 the great majority of the mutants isolated following nitrous acid treatment are revertible to wild type by base analogues, and so are interpreted as transition mutants, a conclusion which fits with the presumed mechanism of action of nitrous acid (cf. Table 6).

Drake found that about half of a sample of UV-induced *rII* mutants were revertible by base analogues but that very few of these were revertible by hydroxylamine; this is the basis for his conclusion, referred to above, that UV induces GC to AT, rather than AT to GC transitions. The other half of the UV mutants, in common with a majority of *rII* mutants of spontaneous origin, are not revertible by base analogues, and would thus seem *not* to be due to transitions. What, then, can be the basis of these mutants? Some of them may be due to what Freese has called *transversions*, that is the substitution of a purine for a pyrimidine and *vice versa*. Freese has argued that a fraction of ethyl-ethanesulphonate (EES) mutants which are revertible by further EES treatment but not by base analogues could be due to transversions, as could some spontaneous mutants which show a similar spectrum of susceptibility to mutagens. There remains, however, a large class of spontaneous and UV-induced mutations which are not revertible by any of the agents we have considered, and which seem more likely to be of the frameshift type, referred to below.

There is one potent class of mutagens, the acridine drugs, of which proflavine was the first to be used which act in quite a special way. Proflavine-induced mutants of T4 are not induced to revert either by base analogues or by EES, and it is now clear that drugs of the acridine class bring about errors in DNA synthesis resulting in the appearance of short repetitions or deletions in the nucleotide sequence. They probably act through the insertion of the polycyclic acridine ring into the stack of base-pairs which runs up the core of the double-stranded DNA molecule, but the precise sequence of events is not definitely known. The most commonly used acridine compounds in current mutation studies are those of the *'acridine-mustard'* class in which the acridine group is linked to a 'mustard' moiety carrying a reactive chlorine. *Frameshift mutations*, as the acridine-induced additions or deletions are called for reasons which will become apparent in the next chapter, are of great importance in connection with studies on the genetic code (see pp. 87).

PROPERTIES OF INDIVIDUAL MUTATIONAL SITES

The first detailed study, using chemical mutagens, of the mutation properties of a large array of sites within a gene, was made by Benzer on the *rII* mutants of bacteriophage T4. The techniques used for mapping these mutants have already been described (cf. Fig. 21).

Within the two genes, A and B, of the *rII* region, Benzer distinguished 308 distinct sites which, between them, mutated over 2400 times in his experiments. The sheer number of sites alone suggests that each site is probably just a singlle nucleotide pair. Using the simple argument outlined on p. 51 one concludes that there are probably of the order of a few thousand nucleotide pairs in the *rII* region of the genetic material, so the number of sites already identified is a substantial part of the theoretical maximum even if every nucleotide pair is a potential site of an observable mutation. In fact, even though there is no reason to doubt that every nucleotide pair can mutate, it does not follow that all mutations will cause an observable change in the properties of the virus, and there is good evidence that a part of the *rII* region can actually be deleted without any drastic effect on its function. Many of the sites have only been observed to mutate once, and it follows that there must also be many which, purely by chance, have not been identified so far but which would be picked up in further experiments. All these arguments encourage the belief that the identified sites *are* single nucleotide pairs, and that it is possible, in principle, to map all the nucleotide pairs in those parts of the genome which have a critical effect on phenotype.

If all the possible mutational sites in a given region could be mapped, and if in addition it were possible, by studies of induction of reversions with specific chemical mutagens, to determine the nature of the nucleotide pair at each site, one would be able to determine the exact base composition and sequence of the region concerned. It may be doubted whether a *complete* determination of the structure of any extensive piece of genetic DNA will ever be made by such methods, but experiments of Benzer and Champe made some progress towards at least a partial solution of the structure of the *rII* region.

We saw earlier how the approach of Strelzoff, which can discriminate mutation by mis-incorporation from mutation by mis-replication, can lead to fairly confident identifications of sites as AT or GC base pairs. A somewhat simpler approach, utilised by Benzer and Champe, is to determine which mutagenic chemicals, if any, are able to induce reversion to wild type in the case of each mutant. For 62 of the 300-odd *rII* sites mapped by Benzer there is evidence from reversion studies permitting base-pair identification. Mutants at each of these 62 sites can be induced to revert by 2-AP or 5-BU, and so would appear to be due to base-pair transitions. Forty-six of the mutants were induced to revert only at relatively low frequency by 5-BU, and were also almost or quite unresponsive to hydroxylamine. Since hydroxylamine seems to react specifically with cytosine under the conditions of these experiments, these 46 mutants were thought likely to have AT at their mutant sites, and thus to correspond to sites carrying GC in the wild type. This result also implies that 5-BU does not cause mutations with high frequency by the mis-replication mechanism, since if it did it would be highly effective at inducing AT to GC transitions. The other 16 mutants were readily induced to revert not only with 2-AP but also with 5-BU and hydroxylamine. These were concluded to have GC base pairs at their mutant sites, and to correspond to AT sites in wild type.

Identifying a base pair at a mutant site does not, of course, answer the question of which base is on which strand of the DNA. Even this problem may be soluble in principle, making use of the fact (cf. p 82) that only one of the two strands of a genetic DNA molecule is used in determining the structure of the 'messenger' RNA which carries the genetic information to the centres of protein synthesis. It would take us too far afield to give the argument here; the interested reader is referred to the paper of Champe and Benzer (1962).

If all sites with the same base pair responded to mutagens in the same way there should be only two, or at most four, classes of site with respect to mutability. What is actually found is very different. In the T4 *rII* region there are a few sites which mutate spontaneously with very high frequency relative to the others. For example, one site in the A gene is represented by some 300 spontaneous mutants, while a great many other sites have only been known to mutate once or twice each, and anything over 20 recurrences at a single site is unusual.

The very highly mutable sites are known as 'hot-spots'—an apt colloquialism which has become established in the scientific literature.

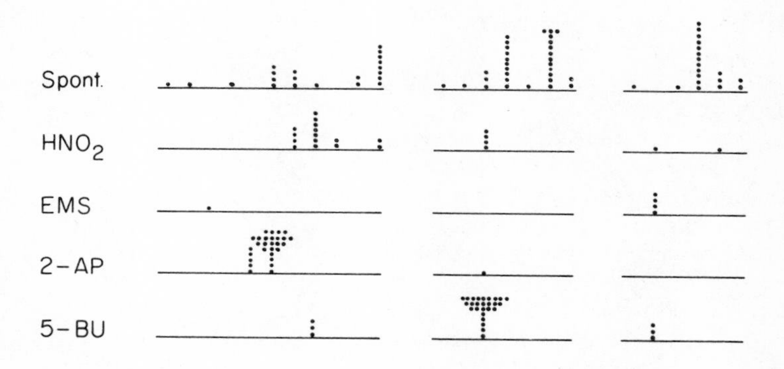

FIGURE 24 Three segments of Benzer's map of the bacteriophage T4 *rII A* and are defined by reference to a standard set of deletions (cf. Fig. 21). Each dot represents a separate mutation; dots on the same vertical line or in the same inverted pyramid indicate mutations at the same genetic site. The regions have been selected to show sites of moderately high spontaneous mutability (though a few much more mutable sites occur in other regions) and other sites which are 'hot-spots' with respect to mutagens. SPONT. = spontaneous; EMS = induced by ethyl methane sulphonate; 2-AP = induced by 2-aminopurine or 2, 6-diamino purine, the two analogues appearently having the same site specificity; 5-BU = induced by 5-bromouracil, 5-bromodeoxyuridine or 5-bromodeoxycytidine, all three of which appear to have the same specifity; HNO_2 = induced by nitrous acid.

There are mutagen-inducible hot-spots as well as spontaneously mutable ones, and it is quite clear, for example, that some GC sites are very much more susceptible to 5-BU than are others.

Because a site is spontaneously mutable it does not follow that it will respond in a spectacular fashion to mutagens; in fact the spontaneous and inducible hot-spots tend to be different. Fig. 24 shows some of the relevant data on the *rII* A gene.

There is no satisfactory general explanation for the existence of mutational hot-spots. Some spontaneously mutable hot-spots probably represent short runs of identical bases, since Yourno has shown that repetition of the same base can predispose the reiterated sequence to a high frequency of spontaneous frame-shift mutation (see p. 87). Base analogue inducible hot-spots are not explained; an influence of adjoining parts of the sequence on the tautomerism of a particular base seems likely, but how such an effect would work is not understood.

5 Gene Action

Thus far we have been concerned with the nature of the genetic material and the spatial relationships of its component elements. In this chapter we turn to the question of how the genes exercise their controlling function in the cell.

ENZYMES

The growth of a micro-organism depends on the synthesis of specific kinds of large molecules, especially proteins, and their organisation into characteristic cell structures. The materials and the chemical energy needed for this macro-molecular synthesis are derived from the transformations of smaller molecules which are themselves synthesised from the components of the growth medium.

Very few of the thousands of different chemical reactions involved would proceed at a significant rate of their own accord. Practically all are promoted by specific protein catalysts, the *enzymes*. Catalysts, which may be defined as substances which facilitate chemical reactions without themselves undergoing chemical change, are well known in inorganic chemistry, and can often be quite simple substances. The special feature of enzymes is the extreme specificity of their action. This is not the place for an account of present knowledge—which is still very imperfect and fragmentary—of the detailed mechanisms of enzyme action. Suffice it to say that very specific complexes are formed between the enzyme and the molecules participating in the reaction (the *substrates*), and that the latter are very much more reactive when complexed with the enzyme than they are ordinarily.

By reason of their large size and structural complexity, proteins as a class show an almost infinite variety of binding reactions, but a particular protein acting as an enzyme binds with high affinity only one or a few kinds of small molecules and will catalyse only reactions involving those molecules. Any living cell contains a vast array of enzymes, each one necessary for the catalysis of a specific reaction, and what the cell does in the way of metabolism and growth is to a large extent determined by the composition of this array. Thus if, without having any experimental evidence, one were asked to hazard a guess as to the way the genetic material exercised its control, it would be natural to suggest that its primary function might be to control the kinds of proteins produced by the cell.

THE GENETIC UNIT OF FUNCTION–THE CISTRON OR GENE

The last two chapters described how it is possible to map the sites of genetic mutations either in a single closed loop, as in some bacteria and bacteriophages, or in several linear linkage groups, as in fungi and higher organisms. Mutations can generally occur anywhere in the genetic map, and from the point of view of formal mapping the arrays of mutant sites are continuous with no clear indication of gaps. The genetic material is, however, clearly segmented from the functional point of view, since mutations having the same kind of phenotypic effect tend to occur within the same short segment.

In order to define more precisely the genetic unit of function it is necessary to have some reliable criterion for distinguishing between different functions. Superficial phenotypes of mutants cannot be relied upon, since similar phenotypes can be due to quite different primary causes. Without undertaking detailed biochemical analysis, the only available criterion is that of the *complementation test,* the principle of which can be most easily illustrated by reference to an actual example from *Neurospora crassa.*

Suppose we have two independently isolated *N. crassa* mutants requiring adenine for growth. By genetic analysis of the cross between them we find that they represent different but quite closely linked mutant sites. In order to find out whether the same function in adenine synthesis has been lost in the two mutants we need to know whether the two mutant genomes are able to complement each other to bring about adenine synthesis when brought together into the same cell.

In *N. crassa* this can easily be determined by formation of a *heterokaryon,* that is to say a mycelium with two different kinds of nuclei intermingled in a common cytoplasm. Heterokaryons are readily formed by any two strains of like mating type and not differing in respect of various known incompatibility factors. When two such strains are inoculated together on an agar medium fusions between growing hyphae, with consequent intermingling of nuclei, occur freely. In the case of our two adenine mutants, a vigorous growth on minimal medium when the two are inoculated together indicates that the two genomes will complement each other, with synthesis of adenine, in a heterokaryon. In this case the simplest explanation is that the two mutations are affecting different synthetic steps in adenine synthesis, although, as we shall see below, there are cases where such a conclusion would not be justified. On the other hand if the two mutants fail to grow in the absence of adenine when inoculated together, and it is known that they are capable of forming a heterokaryon (a point which is easily checked), the inference is that they are defective in the same function in adenine synthesis.

Which result is actually found depends on which two mutants we use. Among the *ad-3* series of mutants in *N. crassa*, all of which are quite closely linked, both situations are found. It is, in fact, possible, as De Serres showed, to divide the whole series into two mutually exclusive groups A and B. Any A mutant will complement in a heterokaryon with any B mutant, but two mutants of the same group show complementation only in isolated cases, and then only to a very limited extent. More detailed mapping of the *ad-3* mutants shows that the A and B mutant sites fall into two distinct and non-overlapping segments of the genetic map, though the two segments are very close to each other. Without the complementation test the existence of these two functionally distinct though related genes would not have been evident.

In 1958 Benzer, in an attempt to cut through the terminological confusion in which genetics then found itself, proposed the use of the term *cistron* for the genetic unit of function. The reason for choosing this term was that it referred to a unit defined by a *cis-trans* comparison.

The heterokaryon test in *N. crassa* consists of making the $a^+b/a\ b^+$ combination of genomes, where a and b represent different mutant sites and a^+ and b^+ their wild type counterparts. This is the *trans* arrangement. The strict control experiment is to make the alternative arrangement of the same genetic components, $a^+b^+/a\ b$, that is, the *cis* arrangement, with the two wild type sites together on the same chromosome. Where the *trans* arrangement gave a mutant phenotype and the *cis* something approaching the wild type, then a and b clearly fall into the same unit of function, or the same *cistron.* If a and b are parts of independently functioning genes then the *trans* and *cis* arrangements should give identical phenotypes.

In practice the *cis* arrangement, involving as it does the isolation of a doubly mutant genome, is often very troublesome to obtain. When one is dealing with mutants giving metabolic defects, the *cis* combination is practically always non-mutant phenotypically, and so a mutant phenotype given by the *trans* combination is usually considered adequate evidence that the mutants concerned are within the same cistron. Thus the *cis-trans* test tends to be replaced by a simple complementation test, complementation in *trans* being *prima facie* evidence that the mutants are in different cistrons, and non-complementation rather conclusive evidence that they are in the same cistron.

At one time many workers wished to use the term *gene* to refer to an indivisible unit of genetic transmission. It now seems clear that, both in micro-organisms and in higher eukaryotes, the genetic material is divisible by recombination right down to the level of nucleotide pairs, and that the smallest unit with any pretensions to specific function is, in fact, the cistron. Consequently the terms gene and cistron tend now to be used as synonyms, and they are so used in this book. The cistron is a discrete segment of genetic material. Mutant sites falling into the same cistron always map within the same short segment of the genome, and are in exclusive occupation of that segment; in other words different cistrons do not overlap, though they can, apparently, be contiguous.

COMPLEMENTATION TESTS IN DIFFERENT ORGANISMS

Complementation tests are possible in most of the organisms and viruses dealt with in this book, but the experimental procedure differs from one organism to another. In higher organisms, and in a few fungi such as ordinary yeast, where there is a prolonged diploid phase of the life cycle, two different genomes can be brought together into the same cell nucleus by an ordinary sexual mating. In *Neurospore crassa*, as we have seen, the ready formation of heterokaryons makes it easy to get two mutant genomes into a common cytoplasm, even though not into the same nucleus. In bacteriophage, as was mentioned in the last chapter, complementation between functionally distinct mutants can be observed following mixed infection of the host bacteria. The genetically useful bacteria, however, present more difficult problems because it is not usually possible to transfer a complete genome from one cell to another, and even the partial diploids which it is possible to form by transduction or (in *E. coli*) by cell conjugation have usually only a very transient existence. Nevertheless, procedures have been worked out for making complementation tests both in *Salmonella typhimurium* and in *E. coli*.

Complementation tests in *S. typhimurium* depend on the phenomenon of *abortive transduction*, first exploited by Ozeki. The most obvious non-lethal result of infection of one *S. typhimurium* mutant by P22 phage grown on another is the formation of non-mutant transduced colonies, which appear in all cases where the strains concerned are mutant at different genetic sites. In addition, many transduction experiments involving pairs of auxotrophic mutants result in the appearance of numerous minute colonies which grow to the point of being barely visible on minimal medium. It has been shown that the growth of each of these minute colonies is due to a single prototrophic cell, which transmits its capacity for growth to only one progeny cell at each division. The explanation of this curious behaviour is that the transducing phage has carried into a recipient cell a complementary fragment of genetic material from the donor, but this fragment has not been integrated into the recipient genome and cannot replicate itself. The presence or absence of minute colonies due to abortive transduction is something which can be easily determined in any experiment in which one auxotroph is infected with phage grown on another; the absence of such colonies may be taken as evidence that no part of the donor genome can carry out the function defective in the recipient, and hence that the two mutations are in the same cistron.

In *E. coli* abortive transduction has not been reported, but in this species special use can be made of the self-replicating fertility (F) factor, mentioned in Chapter 3. Occasionally pieces of the bacterial chromosome become attached to an F factor in its autonomous phase, and the chromosome fragment then becomes endowed with the ability to replicate, along with the F factor, independently of the chromosome. Such an augmented F is known as an F-prime (F′) factor, and it can be transmitted in infective fashion from cell to cell just like an ordinary F. Thus any piece of *E. coli* genome incorporated in an F′ factor can be introduced into any other *E. coli* strain, and its ability to complement other genomes can be determined. This use of F′ factors will be referred to again in the next chapter, and the properties of extra-chromosomal factors of this general sort are discussed in Chapter 7.

ONE GENE–ONE ENZYME

There is now a very considerable body of evidence to support the generalisation that each cistron determines the structure of one polypeptide chain. The one cistron (or one gene)–one polypeptide chain hypothesis is a reformulation of the one gene–one enzyme hypothesis put forward by Beadle and Tatum and their school on the basis of their pioneering work on auxotrophic mutants of *Neurospora crassa.* The reformulation became necessary with the demonstration that some enzyme proteins consist of more than one kind of polypeptide chain, and that, in such cases, the different chains are under the control of different cistrons.

The evidence we shall consider here will be derived from just two classical series of experiments, one on histidine auxotrophs of *Salmonella typhimurium,* and the other on tryptophan auxotrophs of *Escherichia coli.*

Histidine mutants of Salmonella

A very large number of auxotrophic mutants responding to the amino acid histidine were isolated in *Salmonella typhimurium* by Demerec, Hartman and their colleagues. Complementation tests, using the abortive transduction method, served to subdivide the mutants into nine cistrons, which were designated A to I. Some intra-cistronic complementation, of the kind to be discussed below, was found within E and B, but otherwise the classification of mutants into cistrons was very clear cut.

A striking feature of these nine cistrons, and one which is found in some but not all other series of cistrons concerned in the synthesis of other amino acids, is that all are closely linked. In fact all nine are sufficiently close for any two mutant sites within them to show a fairly

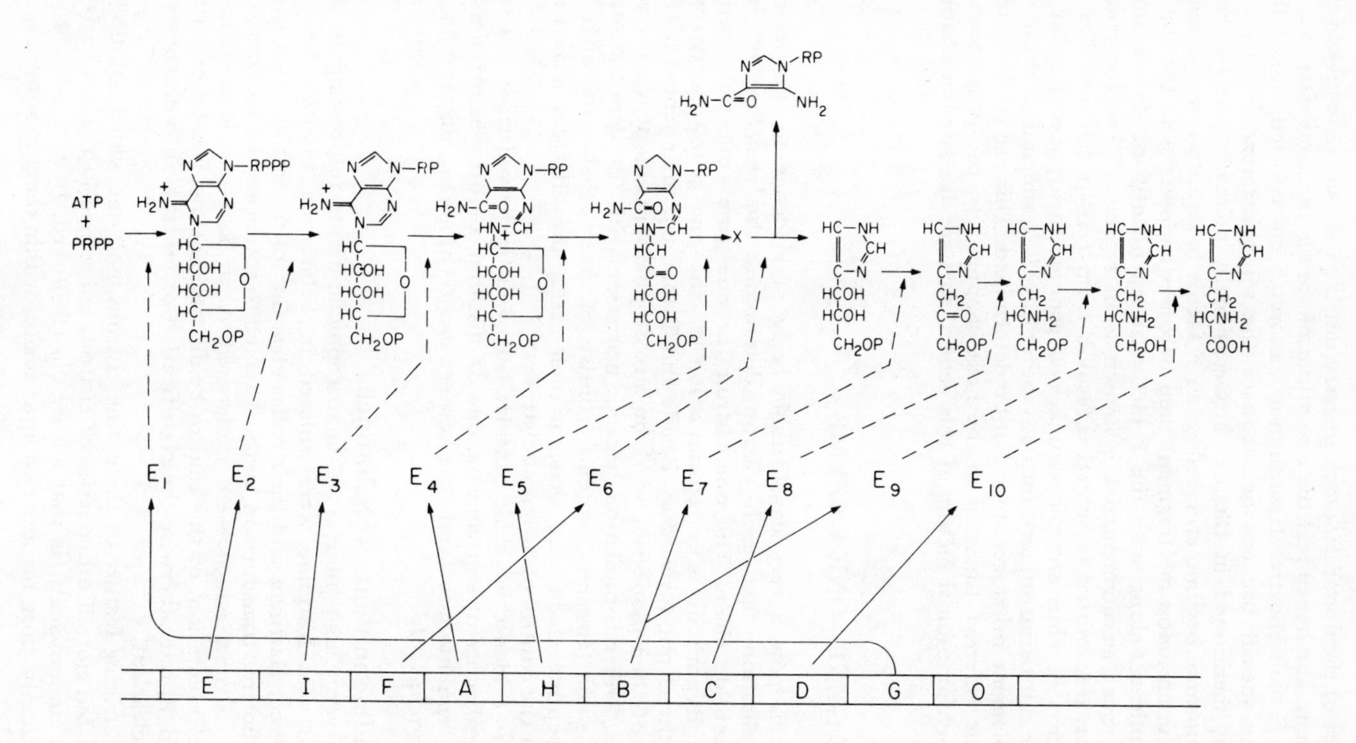
ATP
+
PRPP
E1
E2
E3
E4
E5
E6
E7
E8
E9
E10
E
I
F
A
H
B
C
D
G
O

high frequency of joint transduction (cf. pp. 39-40). Moreover, the relatively short genetic segment within which all the cistrons fall seems to be *exclusively* concerned with histidine synthesis; the histidine cistrons fill the segment, with the sites at the right-hand end of one coming close to the sites at the left-hand end of the next. By transduction analysis of the type described in Chapter 3 the linear order of the cistrons has been determined (Fig. 25).

Through the work of Ames and others the chemical pathway of histidine biosynthesis is now known. The first reaction specifically concerned with histidine synthesis is between phosphoribosyl pyrophosphate and adenosine triphosphate, both of which compounds are involved in the synthesis of many other compounds besides histidine. Following this initial step there are ten other sequential reactions leading to the formation of histidine. Each of these steps is catalysed by an enzyme, and there seem to be eight different enzymes of which one catalyses two different steps.

Ames tested representative mutants of each of the different cistrons for their ability to synthesise the known enzymes of histidine synthesis. For this purpose the mutants were grown on medium supplemented with a suboptimal concentration of histidine, since it is under conditions of histidine starvation that the histidine-synthesising enzymes tend to be produced in the largest amounts (see the next chapter). It turned out that each cistron was specifically concerned with just one of the enzymes; each mutant (with a few exceptions which are relevant to the argument in the next chapter) was unable to make one enzyme, but was able to make all the others. As might have been expected, all the mutants falling within the same cistron were deficient in the same enzyme. The whole situation is summarised in Fig. 25.

There are some important features of this work which at once serve to complicate the picture and make it very much more interesting from the point of view of the *control* of gene activity. These aspects will be considered in the next chapter. In the present context the point to be emphasised is the support given to the one gene—one enzyme hypothesis.

FIGURE 25 Map of the *his* region of *Salmonella typhimurium*, showing the order of the genes and the nature of the reaction catalysed by the enzyme controlled by each gene (after Ames and Hartman, 1963, slightly up-dated). Mutants were assigned to different genes (cistrons) by complementation tests using abortive transduction, and the genes were ordered by transduction analysis. Mutation in gene B causes the loss of enzyme activities E6 and E8, and these seem to be functions of the same protein. The reaction sequence starts with adenosine triphosphate and phosphoribose pyrophosphate at the left and finishes with histidine on the right. Chemical abbreviations: R = ribose residue, P = phosphate residue. Unspecified angles in ring compounds are carbon atoms with hydrogen atoms to saturate the unfilled valencies. O is the operator region (see Chapter 6).

There are now a great many examples all of which point to the one gene—one enzyme hypothesis being true in some sense. Most of these do not give information on exactly *how* the gene is involved in enzyme formation. What is observed is that a particular class of mutants lacks a particular enzyme activity. Is this because the gene, or cistron, is necessary for determining the specific *structure* of the enzyme protein? Or does the mutation cause, in some way, the switching off, so to speak, of the formation of a protein whose structure is determined independently of the gene in question? There is, in fact, evidence that both explanations can apply in different cases, but the most commonly isolated auxotrophic mutants, including the histidine auxotrophs of *S. typhimurium*, are due to mutations in genes with a structure-determining rather than a 'switch' function. For the clearest possible demonstration of genetic determination of enzyme structure we will turn to another class of auxotrophs, the tryptophan requiring mutants of *E. coli*.

Tryptophan mutants of Escherichia coli

The basis for the classification of *E. coli* tryptophan mutants into four groups has been explained in Chapter 3 (see Table 4, p. 40). By far the greatest amount of work has been done on the *trp A* and *trp B* groups, both of which are deficient in the enzyme tryptophan synthetase. This enzyme catalyses the reaction:

indoleglycerol phosphate + serine → tryptophan + triose phosphate (1)

Yanofsky and Crawford purified this enzyme and discovered that it consisted of two components, A and B, which are normally bound together but can easily be separated from each other. Each consists of a single kind of polypeptide chain, α in A and β in B. The separated A component contains one α chain, but there are two α's in the intact enzyme; the separated B component is a dimer containing two β chains and again the intact protein contains two such chains. Neither the A nor the B component can catalyse reaction (1) in the absence of the other, though A has a trace of activity for the reaction:

indoleglycerol phosphate → indole + triose phosphate (2)

while B is somewhat active in catalysing:

indole + serine → tryptophan (3)

Although the partial reactions (2) and (3) can be demonstrated in the test tube, it is unlikely that they are significant in the intact organism. The physiologically important reaction is (1) in which indole is not released as a free intermediate (cf. Table 4). Thus A and B act together as a unit, and their complex can be regarded as a single enzyme from the functional point of view.

The separated A and B components react readily to form the active complex when mixed together. Thus an extract of any mutant producing good A component, but defective in B component, will react with an extract of any mutant with good B component and defective A component to form tryptophan synthetase of the normal type. Using this convenient *in vitro* complementation test, it was shown that mutants of the *trp A* group were defective in the A protein, and that mutants of the *trp B* group were defective in the B protein. By transduction analysis of the type already described on pp. 38-41, it was shown that the two groups of mutants occupied distinct though adjacent genetic segments, which must be regarded as separate genes or cistrons. Thus this situation is one which is consistent with the concept of one gene—one polypeptide chain, rather than one gene—one enzyme.

Since both the α and β chains can be isolated in pure form the way is open for determining the nature of the defects in the proteins brought about by mutations in the corresponding genes. Most of the mutants produce one or other protein in defective form, rather than failing to make it altogether. Yanofsky, working on the α-chain, and Crawford, working on the β-chain, have both demonstrated a variety of subtle effects of mutation on enzymic activity, and these effects are no doubt due to changes in molecular structure. However, without waiting for a detailed chemical explanation of enzyme activity, Yanofsky concentrated on direct chemical analysis of a number of mutant varieties of the α-chain.

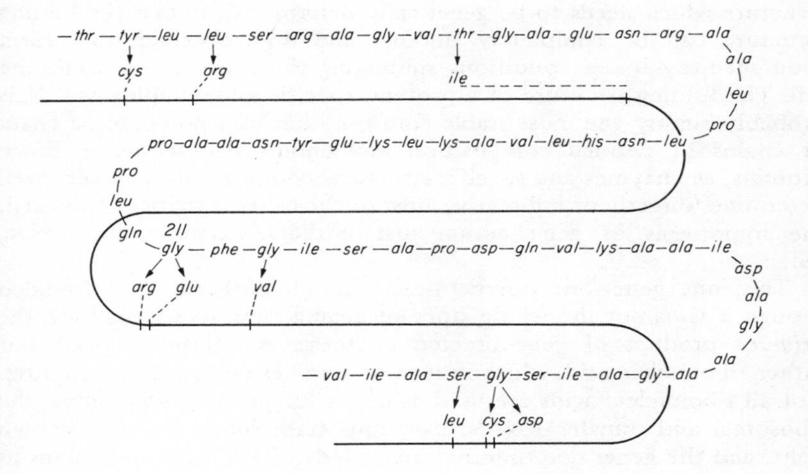

FIGURE 26 Comparison of the positions of mutant sites in the *E. coli* 'A' gene and the corresponding amino acid substitutions in the polypeptide chain of the tryptophan synthetase α subunit. The amino-acid substitutions are connected to the corresponding sites on the genetic map by dotted lines. The mutant sites on the genetic map (solid line) are spaced according to the genetic map distance, proportional to recombination frequency in transduction experiments. Only a portion of the whole polypeptide chain is shown. Amino acids are identified by the first three letters of their name in each case, except for ile = isoleucine, gln = glutamine and asn = asparagine. After Yanofsky *et al.* (1964), with some later amendments.

The results were beautifully simple and clear-cut. Yanofsky found that single mutations bring about *single amino acid substitutions* in the polypeptide chain. At the same time, mapping by transduction analysis made it possible to determine the linear order of the mutant sites in the A gene, and thus to compare this order with the positions in the polypeptide chain of the corresponding amino acid changes. The number of mutants for which the necessary chemical analysis has been completed is now quite large, and the results indicate strongly that the linear structure of the gene (presumably the DNA base-pair sequence) forms a code specifying the linear order of the amino acid residues in the polypeptide chain. Not only does the order of mutant sites correspond to the order of the amino acid changes, but the relative spacing of the sites on the genetic map agrees well with the polypeptide structure. Fig. 26 summarises some of the results. As a result of this study, and others leading to similar conclusions, it seems certain that the function of at any rate a large class of genes, is to determine the amino acid sequences of specific polypeptide chains. Such a function might seem an inadequate basis for a comprehensive genetic control of the cell. However, there is good evidence from studies on a number of proteins that the amino acid sequence is the only aspect of protein structure which needs to be genetically determined, in that the folding structure can be completely undone and then observed to reform spontaneously under conditions simulating those prevailing inside the cell. The folding structure of a protein, specific and complex as it is, is probably simply the most stable conformation of a polypeptide chain or chains of defined composition and amino acid sequence. Since proteins, as enzymes and specific structural components, may very well determine, directly or indirectly, most of the other activities of the cell, the hypothesis of gene action just outlined seems quite reasonable.

The 'one gene—one polypeptide chain' hypothesis, well founded though it is, is not the whole story of gene action. As we shall see, the *primary* product of gene-directed synthesis is not polypeptide but rather ribonucleic acid, which in turn determines polypeptide structure. Not all ribonucleic acids are used as codes for polypeptides; some, the ribosomal and transfer RNA's, have important functions in their own right, and the genes determining these kinds of RNA are exceptions to the 'polypeptide' generalisation of gene action. Evidence for the existence of genes determining transfer RNA structure is touched upon later in this chapter.

INTRA-GENIC COMPLEMENTATION

At this point it seems desirable to turn aside from the main argument to consider a phenomenon which, at first sight, seems to call into question the validity of defining genes on the basis of complementation tests. It often happens that a series of mutants of related phenotype show no complementation in any combination. One thus has no hesitation in assigning them to the same cistron or gene. Sometimes, however, as the number of mutants in the series is increased, a few pairwise combinations are found which show a limited degree of complementation, producing a relatively small amount of the enzyme activity which is absent, or present only in traces, in the mutants grown separately. This kind of limited complementation cannot be used to define genetic segments of independent function, because it is usually found that each two partially complementing mutants agree in not showing complementation with most other mutants of the series. Recent studies on the biochemical basis of such situations suggest that intragenic complementation, as it may be called, does not necessitate any revision of the one gene–one polypeptide chain concept. The generally accepted explanation is as follows.

A great many enzymes, probably a large majority, are aggregates or *oligomers,* consisting of a fixed number of polypeptide chains. An enzyme oligomer may contain just one kind of chain or more than one, but in any case there are usually two or more copies of each kind. It is the individual chain, rather than the complete oligomer, which is the product of a single gene. Thus, when two different mutant varieties of the polypeptide are being produced simultaneously in the same cell a proportion of *mixed oligomers* will probably be formed. Where two different defective polypeptide chains are present a mixed oligomer has sometimes at least partial enzyme activity though both kinds of pure oligomer are inactive. The most usual reason for this is thought to be that an abnormal secondary or tertiary structure, which may be the reason for the inactivity of each kind of pure mutant oligomer, is corrected by association of two mutant polypeptide chains with *different* defects; the faulty section of each chain may be stabilised in the correct folding pattern by the corresponding 'good' section of the other. This type of explanation implies that enzymes formed as a result of complementation between mutants in the same gene should still not be completely normal. It might be expected, for instance, that they would be rather unstable and/or of abnormally low activity. This does, indeed, seem commonly to be the case.

BIOCHEMICAL EVIDENCE ON THE MECHANISM OF SYNTHESIS OF RNA AND PROTEINS

The conclusion at which we have arrived is that the linear sequence of genetic sites in the gene forms a code for the linear sequence of amino acids in the corresponding polypeptide chain. Before going on to discuss further genetic evidence on the nature of this code and how it works it may be helpful to outline very briefly our present rather full understanding of the biochemical mechanism of protein synthesis.

Transcription of DNA into RNA

Although DNA must, as the genetic material, exercise the ultimate control over protein structure it plays no *direct* role itself in protein synthesis. The sole known biochemical function of DNA, apart from directing its own replication, is to act as template for RNA synthesis. The process of DNA-directed RNA synthesis is conveniently referred to as *transcription.* An enzyme, RNA polymerase, will synthesise RNA molecules complementary to one strand of a DNA duplex by a mechanism similar to that of DNA synthesis but with uracil substituting for thymine as the pairing partner of adenine and with the ribonucleoside instead of the deoxyribonucleoside triphosphates as precursors. Thus the genetic code embodied in the base sequence of a DNA strand is transcribed into a different but closely related language—the base sequence of a complementary strand of RNA. It seems that transcription cannot start just anywhere on the DNA but rather has fixed starting and stopping places, and that only one strand of any particular segment of DNA is normally transcribed. RNA polymerase recognises transcription initiation points through a component of its structure called *sigma.* The *sigma* factor can be dissociated from the rest of the enzyme, leaving a polymerase which initiates transcription indiscriminately. It has been supposed that there may be a set of interchangeable *sigma* factors which promote the transcription of different sets of genes, but evidence in favour of this attractive idea is so far rather meagre.

A specific *transcription-termination* factor, named *rho*, has also been identified in *E. coli*, but the way in which it works is not understood; presumably it must have specificity both for the RNA polymerase and for the appropriate termination points on the DNA.

Ribosomes

Three different classes of RNA molecules transcribes from DNA are involved in protein synthesis. Firstly, there is *ribosomal RNA*. All protein synthesis occurs in association with particles called *ribosomes.* The best studied ribosomes are those of *Escherichia coli* which each consist of two subunits, a large one containing an RNA molecule about 3000 nucleotides long and one molecule each of thirty four different proteins (called the 50s particle after its velocity of sedimentation in the ultracentrifuge) and a smaller one with an RNA molecule of about 1500 nucleotides and one each of about twenty one other protein molecules (the 30s particle). The 50s particles also each contain a small additional RNA molecule only about 100 nucleotides long (5s RNA). These three kinds of ribosomal RNA are transcribed from specific segments of DNA, or ribosomal RNA genes, which appear to be present in the chromosome in multiple copies and together make up about 0.3% of the DNA. Ribosomes of eukaryotic organisms are very similar in general to those of bacteria except that their subunits are slightly larger and their functioning is inhibited by a different set of drugs (e.g. cycloheximide for eukaryote ribosomes and streptomycin or chloramphenicol for prokaryote ribosomes).

Messenger RNA

Cell-free preparations of ribosomes will carry out at least a limited synthesis of polypeptide chains, given the amino acid precursors and several essential co-factors including adenosine triphosphate, guanosine triphosphate and magnesium ions, *provided* that two other kinds of RNA are present in the system. One of these is the so-called *messenger* RNA (mRNA), to which ribosomes attach, several to each mRNA molecule, to form *polyribosomes* (or polysomes). It is the mRNA which 'programmes' a ribosome to synthesise a specific kind of polypeptide chain. A given mRNA molecule is, in fact, a transcript of the specific gene coding for that polypeptide chain (or sometimes, as we shall see in the next chapter, of a cluster of contiguous genes coding for several polypeptides). The mechanism used by the ribosome for *translating* the RNA message into the specific amino-acid sequence of the polypeptide involves the third class of RNA molecules, the *transfer* RNA's or tRNA's.

Transfer RNA

The tRNA's are relatively small as RNA molecules go, containing some 80 to 90 nucleotides; owing to the mutual affinity, through runs of complementary base pairs, of different parts of a tRNA chain, the molecule has a complex secondary structure consisting of a series of hairpin loops. There are probably in the region of 50 different kinds of tRNA molecule in a cell, and each one acts as a carrier, or adaptor, for one amino acid. A series of specific amino-acid activating enzymes, otherwise known as aminoacyl-tRNA synthetases, condense the carboxyl group of the amino acid with the free 3′-hydroxyl group of the terminal nucleoside of the tRNA chain, which is always adenosine. Each kind of tRNA has in its structure, in an exposed position at the turn of one of the hairpin loops, a specific sequence of three nucleotides which allows the mRNA to recognise that tRNA molecule as carrying a particular amino acid. This identifying triplet is called the *anticodon* because it corresponds to a complementary sequence of three in the mRNA which is the coding unit or *codon* for the amino acid carried by the tRNA.

The process of polypeptide synthesis

We are now in a position to describe the course of events in a bacterial cell as successive amino-acid residues are added to a growing polypeptide chain. The ribosome possesses two sites which can bind to two adjacent codons of the mRNA. These are called the 'peptidyl' (P) and 'aminoacyl' (A) sites respectively. The initiation of peptide synthesis depends on the formation of an initiation complex comprising a ribosome[1], a molecule of a special kind of tRNA carrying the

[1] Initially it is the 30s ribosomal subunit which is involved.

substituted amino acid N-formylmethionine and a molecule of mRNA. The mRNA is bound to the N-formylmethionyl-tRNA anticodon, and also to the P site of the ribosome, by the specific nucleotide sequence AUG which is the codon for methionine. The unoccupied A site of the ribosome is now aligned with the *next* codon of the messenger and will bind the specific kind of aminoacyl-tRNA with the corresponding anticodon. Thus there may be, say, alanyl-tRNA bound at the A site and N-formylmethionyl-tRNA bound at the P site. There now occurs a complex reaction as a result of which the N-formylmethionyl residue is transferred from its tRNA to form a peptide bond with the alanine (to

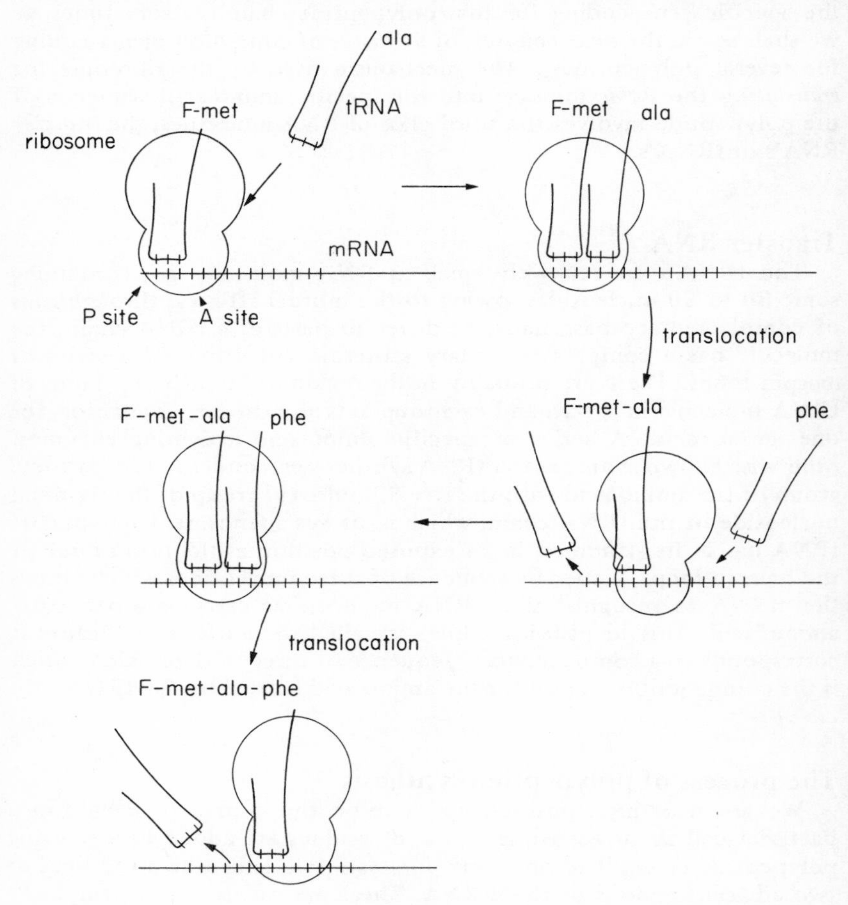

FIGURE 27 The synthesis of a polypeptide chain by a ribosome programmed by a mRNA molecule. The tRNA molecules, which act as keys for translation of the code, are shown diagrammatically as having only one hairpin loop (the anticodon loop) in their structure instead of the several loops which they actually possess. For further explanation see text.

form N-formylmethionyl alanyl-tRNA); and the alanine-specific tRNA, now bearing this dipeptide, is transferred (translocated) to the P site, together with its associated mRNA codon, in place of the N-formylmethionyl tRNA formerly at that site. The A site, now once again vacant, is free to attract a further aminoacyl-tRNA with an anticodon corresponding to the third codon of the messenger. A further peptide bond is now formed, the tRNA at the A site, now carrying a tripeptide, is transferred to the P site, and the process continues. With each transfer of tRNA from one ribosomal site to the other the messenger moves too, by one codon. Thus the ribosome travels codon by codon along the messenger with addition to successive amino acid residues to the growing polypeptide chain. Termination of chain synthesis occurs, and the ribosome dissociates from the mRNA, when the ribosome encounters one of the three chain-terminating codons (UAG, UGA and UAA) which code for no amino acid. The whole process is summarised in Fig. 27.

The code

It is not possible in this brief account to describe the beautiful biochemical experiments which establish the mechanism just outlined. The two main lines of evidence which have led to a complete elucidation of the genetic code may, however, be mentioned. The first depends on the finding of Nirenberg that the binding of each particular kind of tRNA to ribosomes can be stabilised by an isolated codon—that is a free trinucleotide—of the appropriate kind. Since synthetic nucleotide chemistry has now been developed to the point where all the possible trinucleotides can be synthesised, the method has permitted the compilation of a complete catalogue of codons. The second method, even more decisive, depended on the synthesis by Khorana and his colleagues of long polynucleotides of defined repeating sequence. These were used as artificial messengers to 'programme' ribosomes in the presence of tRNA and other necessary factors in a cell-free system, and the amino-acid sequences of the polypeptide chain synthesised were determined. The results were fully consistent with the codon assignments obtained from Nirenberg's method, which are summarised in the table on page ii (Frontispiece).

Although it was first worked out using *E. coli* ribosomes and tRNA, the code seems to apply to all organisms. There is a minor difference, apparently, between bacterial and mammalian cells in chain initiation; although in both kinds of system AUG is the initiating codon and a special tRNA is used for initiation (different from that used for inserting methionine into internal positions in a chain), the initiating methionyl-tRNA is apparently not formylated in mammalian cells. The reason why all polypeptide chains do not begin with N-terminal methionine (a substantial minority do so in bacteria and a smaller minority in eukaryotes) appears to be that the terminal methionine is regularly removed from many nascent chains after initiation has been achieved.

GENETIC CONFIRMATION OF THE CODE

The above account of protein synthesis and its direction by DNA via RNA rests on experiments with artificial cell-free systems. What assurance do we have that the mechanisms in the living cell are similar? The best answers to this question are provided by studies on mutants.

Recombination within a codon

The unit within the messenger RNA which codes for just one amino acid is called a codon. The fact that single-site mutations bring about single amino-acid substitutions shows that the unit of mutation is no larger than a single codon. It also shows that, if the codon is a triplet of bases or base pairs, adjacent codons are non-overlapping—that is that each base is a member of only one coding triplet. Assuming a triplet code, and that genetic recombination can occur between any two adjacent base pairs (cf. p. 51), it would be expected that separable mutational sites should sometimes be detectable within a single codon. Yanofsky has produced evidence that this is indeed the case.

Fig. 28 shows that two different mutants have been isolated in which the tryptophan synthetase A protein is altered in the same amino

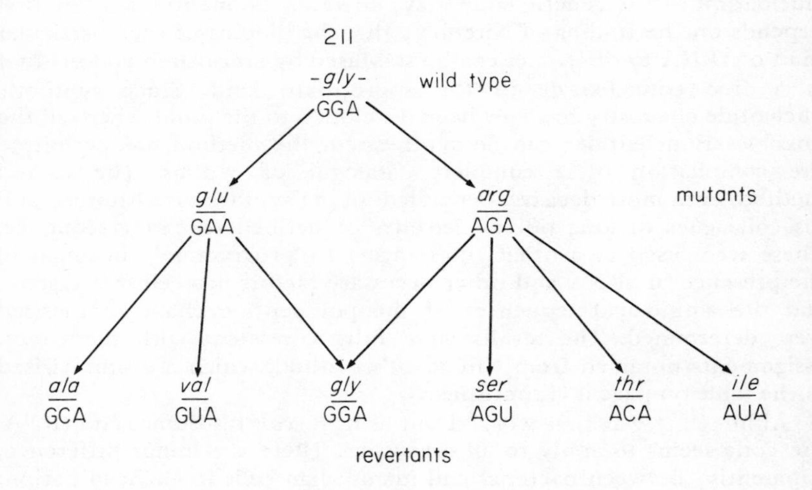

FIGURE 28 Mutation within a single codon specifying the amino-acid residue in position 211 of the *E.coli* tryptophan synthetase α-chain. Beneath each amino acid is the triplet codon (A = adenine, U - uracil, G = guanine, C = cytosine) which, on the basis of experiments outlined on p. 85, is known to code for that amino acid. These are RNA codons and each RNA base corresponds to one member of a DNA base-pair. Single base-pair substitutions in the DNA bring about single base substitutions in the corresponding RNA. There are some alternative possibilities in the last position of the codon in each case (cf. Frontispiece, p. ii). From Yanofsky *et al.*, *Cold Spring Harb. Symp. Quant. Biol.*, **31**, 151 (1966).

acid residue. In one case glutamic acid and in the other case arginine has been substituted for the glycine residue normally present in position 211, numbering from the amino-terminal end of the α-chain. These two mutations map genetically at two different though very closely linked sites. In other words, transducing phage grown on the 'arginine' mutant and used to infect the 'glutamic acid' mutant, or vice versa, bring about the formation, at very low frequency, of recombinant bacteria with glycine restored at the position in question. Thus it is demonstrated that the 'good' parts of two mutant codons can be recombined to give a wild type codon, and hence that the codon must correspond to at least two mutational sites and, by implication, at least two base pairs in DNA.

Successive mutations in the same codon

The work of Yanofsky and his colleagues on mutational changes in the tryptophan synthetase α-chain yielded much confirmatory information on the relationships between the codons for different amino acids. For example, the two mutants mentioned in the last paragraphs, with glutamic acid and arginine respectively substituted for the glycine present at position 211 in the wild type, can each be further mutated and derivatives selected which have restored ability to grow on medium lacking tryptophan. Some of the revertants have glycine restored in position 211, but others have a different amino acid which is apparently a sufficiently good substitute for glycine to permit the enzyme to be normally, or nearly normally, active. Different revertants from the 'glutamic acid' mutant had glycine, alanine and valine in position 211, while different ones from the 'arginine' mutant had glycine, serine, threonine and isoleucine. Fig. 28 shows how our present knowledge of the genetic code—derived from the independent biochemical evidence mentioned above—is entirely consistent with all these mutations being single base-pair substitutions in the DNA.

Evidence from frameshift mutations

The most impressive evidence of a purely genetic character on the nature of the code was due to experiments in which Crick and co-workers used the classical *rII* series of mutants of phage T4, which had already been so profitably exploited by Benzer.

The starting point of the investigation was an *rIIB* mutant which had been induced by proflavine. Proflavine-induced mutants are generally not revertible by base analogues, though they may be so by further proflavine treatment, and they usually show total loss of gene function. As we shall see, they appear to represent small DNA deletions or repetitions, involving one or a few base-pairs, rather than base-pair substitutions.

Crick isolated a series of spontaneous revertants of this mutant which had regained the ability (lost in *rII* mutants) to grow on *E. coli* strain K (λ). Genetic analysis showed, however, that each of these revertants still carried the original mutation together with a second closely linked 'suppressor' mutation, also within the *rIIB* cistron. The suppressors, when separated by genetic crossing-over, proved to give the typical *rII* phenotype (large plaques on B, no growth on K) when present as single mutations. Thus it appeared that the suppression was mutual; two *rIIB* mutations when present together somehow cancelled each other out so far as gene action was concerned. Several isolated suppressor mutations were again used in reversion experiments and numerous second-order suppressors, each nullifying the effect of the primary suppressor, were obtained, and shown to map at yet other closely linked sites within *rIIB*. These, again, turned out to give the typical *rII* phenotype when separated as single mutations.

Still more remarkable were the results of experiments in which the various first- and second-order suppressors were combined together by selecting double- and triple-mutant recombinant phages from appropriate crosses. Calling the original mutant 'minus', the first-order suppressor mutations 'plus' and the second-order suppressors 'minus' again, it was found that most plus-minus double mutants gave the 'wild' phenotype—i.e. pluses and minuses usually cancelled each others' effects. Plus-plus or minus-minus combinations were always mutant. However, several triple mutants of the type plus-plus-plus or minus-minus-minus were phenotypically wild type.

These observations found a beautiful explanation in terms of shifts of a reading-frame for a code written in three-letter symbols (*codons*). Supposing the code for the gene product has to be decoded from one end, reading off in threes from a fixed point, a deletion or addition of a letter, or a short number of letters other than three or a multiple of three, will upset the reading of all the following codons up to the end of the message. A second nearby addition or deletion will restore the reading frame to its correct phase if it either cancels out the first (e.g. minus-one following plus-one) or, in combination with the first, makes a change of three or a multiple of three (e.g. minus-two following minus-one). When this happens the decoding of the remaining message will proceed normally following a region of disturbance which, if it is short, may not upset the functioning of the protein product too much. Similarly three plus or three minus mutations in sequence will, between them, leave the reading in phase, but two of the same sign can never do so. The fact that it was *three* mutations of like sign that restored normal reading was powerful evidence for a three-letter code.

Double-frameshift analysis

Crick's deductions about frameshift and double-frameshift mutations in the *rII* gene of T4 bacteriophage were later fully borne out by studies of genes, in T4, *E. coli* and *Salmonella typhimurium*, of which the protein products can be isolated and characterised chemically. In such cases it is possible to determine the altered sequence of amino acids corresponding to the interval between two frameshifts. By comparison of the normal with the double-frameshift sequence it becomes apparent that only one unique sequence of codons, or at most a very small number of alternatives, *could* yield codons for the altered

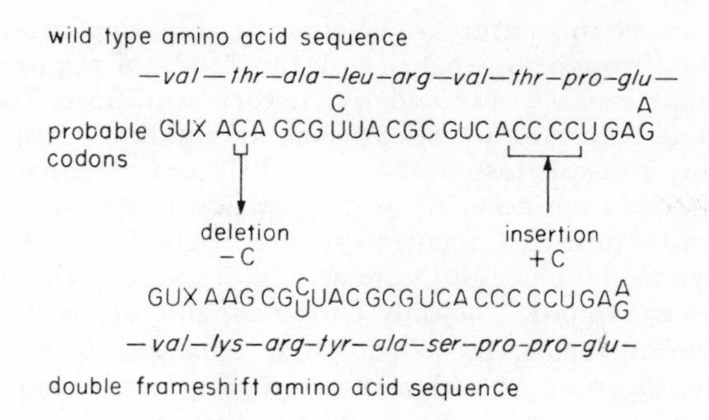

FIGURE 29 Result of an analysis by Yourno (*J.Mol.Biol.*, **48**, 437 (1970), of the relevant part of an altered histidinol dehydrogenase resulting from two compensating frameshift mutations in *hisD* of *Salmonella typhimurium*. The first mutation eliminated histidinol dehydrogenase (controlled by *hisD*) and exerted a strong polar effect on the expression of the more distal genes of the *his* operon. The compensating mutation restored histidinol dehydrogenase activity and relieved the polar effect. In the codon assignments X may be any of the four RNA bases; base symbols written one above the other are indistinguishable alternatives.

sequence of amino acids through a frameshift. Fig. 29 shows an example of such an analysis. This kind of approach is the only one available (short of a direct chemical analysis of the gene, which is as yet possible only in a few cases) for determining which codons are actually used in the intact organism as opposed to the various artificial test-tube systems.

Chain-terminating mutants

Mutants which have lost an enzyme activity often have replacement of essential amino acids due to the conversion of one amino-acid codon to another. Most amino-acid replacements, however, do not bring anything approaching complete loss of the protein function, and many, perhaps a majority, of auxotrophic and other functionally defective mutants are ones which fail to complete the synthesis of the polypeptide chain in question, rather than producing a complete chain with one 'wrong' amino acid. Premature termination of polypeptide chain synthesis results from a mutation of a 'sense' codon, that is one coding for an amino acid, to a 'nonsense' codon, that is one of the three UAG, UAA and UGA which code for no amino acid. Clearly there are many codons which can mutate to one of these three by a single base change (cf. Frontispiece) and the effect, a shortened polypeptide, will in the great majority of cases be incompatible with the formation of a functional protein.

In a few cases the nature of a chain-terminating mutant has been demonstrated through an analysis of the kinds of restoration of gene function which can be obtained by further mutation. For example, a chain-terminated mutant of *E. coli* lacking alkaline phosphatase could regain alkaline phosphatase activity by different mutations restoring a variety of *different* amino acids in the position in the polypeptide chain at which termination had occurred in the original mutant. In different revertants, lysine, leucine, glutamic acid, glutamine, serine, tyrosine and tryptophan were found. The only codon capable of yielding codons for all these amino acids by single base changes is UAG. Another chain-termination mutant which reverted to give the same spectrum of amino acids with the exception of tryptophan, which was never found, was taken to have UAA as the mutant codon.

A very important part in the analysis of chain-terminating mutations was played by the discovery of genes in some strains of bacteria which will mask, or *suppress*, the effects of such mutations. In *E. coli*, for example, the so-called 'amber'-suppressors will over-ride chain termination by the UAG codon, while another class will suppress either UAA ('ochre') or UAG ('amber') mutations. Thus some classes of chain-terminating mutants have little or no effect in strains carrying one of the appropriate suppressors, while leading to complete loss of gene function in strains lacking a suppressor.

The suppression of chain-terminating mutants is now explained as a change in the translation mechanism which permits the reading of the 'nonsense' codon as if it were some kind of 'sense'. Thus one kind of 'amber' suppressor results in the translation of UAG as tyrosine. This is due to a change in the anticodon of a species of tyrosine-specific tRNA from GUA, which pairs with the tyrosine codons UAC or UAU, to CUA, which pairs with UAG. The suppressor genes, in fact, are genes from which tRNA is transcribed, and they exemplify the point made on p. 80 that not all genes code for polypeptides.

Normal chain termination

The fact that the UAA, UAG and UGA codons cause premature chain termination in some mutants does not, of course, necessarily mean that they are the *normal* chain-terminating signals in the wild type. There is, indeed, a good reason for doubting whether the 'amber' codon UAG plays any normal role in termination. Some amber-specific suppressors in *Escherichia coli* can cause the reading of mutant UAG codons with high efficiency so that up to 80% or so of the gene product can be restored without much impairing viability and growth. Were UAG responsible for normal termination of many indispensable polypeptide chains, one would expect such a high rate of failure to terminate to be lethal, since an arbitrary extension of a chain at the C-terminal end would be likely to impair its normal folding and function. On the other hand the *E. coli* suppressors which will act *either* on 'amber', UAG, *or* on 'ochre', UAA, are never very efficient in causing the translation of these codons and, even so, cause a marked reduction in the growth rate of cells carrying them. The implication is that UAA may well be a normally used chain-terminating signal, and other evidence suggests that UGA may be also.

The most direct approach to this question is through the analysis of the nucleotide sequence of the RNA of certain simple RNA bacteriophages such as Qβ and MS2, which code for as few as three proteins: a single protein which forms the head which encapsulates the infectious, RNA, an RNA-replicating enzyme, and a third protein of less well-defined function. Nucleotide sequences of substantial sections of these RNA's, some of which overlap ends of genes for defined polypeptide chains, have been determined, and the UAA codon does indeed occur in the terminating position in at least some cases.

Mutations in initiating codons

Rather surprisingly, no bacterial mutant failing to produce a particular polypeptide chain has been shown to have undergone a change in the initiating codon. However, mutants of this type have been identified in the cy_1, gene of yeast, *Saccharomyces cerevisiae,* coding for the respiratory protein pigment cytochrome c. As in the case of premature-termination mutants, the failure-of-initiation mutants have been characterised by analysis of revertants; these revertants have, in some cases, acquired a new initiation codon a little way to one side or the other of the non-functional mutant initiator, giving either a slightly longer or a slightly shorter chain. In the former case the longer chain includes, at the N-terminal end, an amino acid translated from a mutant derivative of AUG—different ones in different mutants. For a fuller account of this work, which has several other features which serve as excellent examples of the kinds of mutational effects on polypeptide chains dealt with in this chapter, the reader is referred to a recent review by Sherman and his colleagues (Stewart *et al.*, 1972—reference on p. 141)

6 Regulation of Gene Action

CONTROLS OF QUANTITY AND CONTROLS OF ACTIVITY–ALLOSTERIC PROTEINS

The first great period of investigations of gene action, which started with Beadle and Tatum's discovery of auxotrophic mutants of micro-organisms, was directed to the problem of the nature of the gene product. As we saw in the last chapter, the *qualitative* aspect of gene action is now broadly understood, following the elucidation of the mechanisms of transcription of DNA to RNA and of translation of RNA nucleotide sequence into polypeptide amino-acid sequence.

Equal in importance to the determination of the structures of the protein products of genes is the purely quantitative control of gene activity. There are two levels at which quantitative control can be exercised. Firstly, the *rate* of protein synthesis can be regulated and, secondly, the *activity* of a given quantity of protein already synthesised may also be subject to regulation.

This chapter will be mainly concerned with the former aspect, but in order to explain the regulatory mechanisms controlling quantity of protein it is necessary at least to mention how regulation of activity can occur.

One of the most important concepts in modern biochemistry is that of *allosteric* proteins. This has become a most complex subject with a vast literature of its own, but here we need only deal with it in the simplest terms. As was mentioned in the last chapter the three-dimensional structure (or *conformation*) of a protein in a given set of conditions is determined by the primary structure, or amino-acid sequence, of its component polypeptide chains. However, many proteins have more than one stable conformation open to them, and the choice between one and another may be critically dependent on conditions, the most important of which is the binding to the protein of smaller molecules. One or another conformation will often be stabilised by such binding, and the presence or absence of the bound molecule, or *ligand*, will then tend to control the transition between the (usually two) conformational states. Thus, in the case of an enzyme, if one conformation is active and the other inactive the ligand will act as an activator or inhibitor depending on which of the two forms it tends to stabilise. Proteins which display the property of ligand-controlled conformational transition are called *allosteric* proteins and the controlling ligands *allosteric effectors*.

A general and possibly universal feature of allosteric proteins is their oligomeric structure, offering multiple sites for ligand binding. The ligand molecules usually act co-operatively in effecting the allosteric transition, with one molecule bound having rather little effect, two together having much more than twice the effect of one and so on for higher numbers. *Co-operativity* means that the relation between effect and ligand concentration is an S-shaped one, so that a major conformation shift may occur over quite a narrow range of ligand concentration.

In many metabolic pathways in micro-organisms the enzyme catalysing the first step has allosteric properties, with the end product of the pathway acting as an effector inhibiting the enzyme activity. Thus, as the end product accumulates, a negative feed-back mechanism operates to slow the entire pathway through inhibition of its first step. Feedback inhibition acts as a rapid fine control of metabolism, while regulation of enzyme synthesis, with which this chapter is mainly concerned, is a slower-acting control more appropriate for adjustment to longer term changes in the environment.

We shall not be further concerned with allosteric effects on enzymes, but later in this chapter we shall see the importance of the allosteric properties of proteins with functions in the regulation of enzyme synthesis.

INDUCTION AND REPRESSION

It is quite clear that, in microbial cells, the quantities of different proteins are stringently controlled so as to keep them in proper balance and to ensure that cell growth occurs with minimum utilisation of scarce nutrients. Bacteria provide many examples of effects of conditions of growth on the relative amounts of different proteins in the cell. These effects are of two kinds.

Firstly, there are many cases of substances, for example sugars of various kinds, which *can* be used by a bacterial species but are not usually present in the environment. In such cases it is generally found that when the substance in question is present in the growth medium the enzymes necessary for its utilisation are produced, often in quite large amounts, while in the absence of the substance these enzymes are hardly produced at all. This phenomenon is known as *enzyme induction*, the *inducer* usually being the substance for whose utilisation the enzymes are required.

The second kind of effect may often be observed when the growth medium is supplied with adequate amounts of a compound which the bacterium can otherwise make for itself. In such cases it is commonly found that the formation of the enzymes, whose function is the synthesis of the compound in question, is *repressed* by the presence of the end-product of their activity. As a result of these two regulatory mechanisms, the organism tends to form the enzymes necessary for the full exploitation of its environment, while avoiding forming those whose presence is not for the moment necessary.

How do induction and repression work? As in so many other cases, our best understanding comes from studies of mutants in which the normal mechanisms have been altered.

THE *LACTOSE* MUTANTS OF ESCHERICHIA COLI

Any account of regulation of gene action is inevitably dominated by the work of Monod, Jacob and their colleagues and followers on mutants of *E. coli* defective in their ability to utilise the sugar lactose as a carbon source. This work has resulted in a remarkably detailed, complete and well-substantiated model for one system of genes; one of the most important questions we shall have to consider is the extent to which the model is generally applicable to other systems.

The structure-determining genes

The starting point for a long series of experiments was the isolation of many lactose-negative mutants. These fall, for the most part, into two quite distinct classes, due to mutation in two adjacent but non-overlapping segments of the genetic map.

The first class, the *Z* mutants, are defective in the enzyme β-galactosidase, the enzyme which hydrolyses lactose to glucose plus galactose. Many of these mutants have been shown to produce structurally abnormal and inactive forms of the enzyme. The second class, the *Y* mutants, are capable of forming normal β-galactosidase but are defective in their ability to take lactose into the cell from the medium.

The ability of the wild type organism to concentrate lactose from the growth medium depends on a protein which binds lactose (or other β-galactosides) with very high affinity and which is a component of the cell membrane. This protein is a representative of a class of proteins involved in transport of small molecules across cell membranes and sometimes called *permeases*. It appears to be absent or defective in Y^- mutants. The β-galactosidase and permease are present only in minute amounts in cells grown on glucose but their synthesis is induced by lactose or several other galactosides.

A mutant with a normal *Z* gene (Z^+) but with a defective permease (Y^-) has some difficulty in making β-galactosidase since this enzyme normally has to be induced by the presence of a derivative of lactose, or of any one of several other galactosides, within the cell. When the medium contains high concentrations of lactose, however, enough will enter the cell by passive diffusion to bring about the induction of normal levels of β-galactosidase. It is possible to show that the permease itself, in cells able to form it, is induced by a derivative of intracellular lactose. Thus even the wild type has difficulty in adapting itself to a very low concentration of lactose as sole carbon source, since it is not until a little lactose has entered by diffusion that the permease is formed, and it is not until a little permease is present that lactose can be concentrated from a very dilute solution. The point to be emphasised here is that the β-galactosidase and the permease are under independent structural control; most permease-deficient mutants are able to make perfectly good β-galactosidase, and *vice versa.*

It later emerged that adjacent to the *Y* gene, on the side farthest from *Z* is another gene *A* responsible for the structure of another and quite distinct enzyme, thiogalactoside transacetylase. This enzyme has no known essential function and mutant strains with the *A* gene partly or wholly deleted can grow on glucose or lactose, but its formation is induced, along with the β-galactosidase and permease, by the same range of β-galactosides as induce these other two proteins.

Remarkably enough, the rates of synthesis of the transacetylase and β-galactosidase show a constant ratio under almost all conditions of induction. The permease is less easy to assay quantitatively (its measurement depends on observing the rate of intracellular accumulation of a non-utilisable galactoside) but all available information is consistent with the synthesis of this protein being geared in constant ratio to that of the other two. Thus there must be some unitary or co-ordinate control of the action of the group of adjacent genes coding for these three proteins. Such a co-ordinated group of genes was named as *operon* by Jacob and Monod. Further support for the operon concept is considered below.

Constitutive mutants–repressor and operator

Although in wild type *E. coli* β-galactosidase and the two other proteins co-ordinated with it are present only in minute concentrations in uninduced cells, many mutants have been isolated which form maximum amounts of all three even in the absence of inducer. These are called *constitutive* mutants and are of two kinds.

The first type of constitutive mutant to be analysed mapped within a segment adjacent to *Z* on the side farthest from *Y* (see Fig. 30). This segment was named *I* for *inducible*, the wild type character modified in the mutant. The characteristic property of the constitutive *I* mutants (I^-) is that they are *recessive*. Whether a mutant is dominant or recessive is a question that can easily be answered in an organism in which a stable diploid or heterokaryotic phase can be obtained. In *E. coli* the problem is a little more difficult, but it can be overcome through the use of F-prime (F′) elements, referred to in the previous chapter, in which bacterial chromosome fragments have become attached to the fertility factor F, thereby acquiring the capacity for independent replication. Starting with an F′ carrying the *lac* region of the bacterial genome (F-*lac*), and introducing it into various *lac* mutant strains, a range of mutant F-*lac* elements can be obtained as a result of exchange of material between F-*lac* and the mutant genomes. By making use of their property of infective inter-cellular transfer, these mutant F-*lac*'s can then be used to show the phenotypic effects of different pairs of mutant *lac* segments present together in the same cell.

By saying that I^- mutants are recessive one means that a cell with two *lac* segments, one I^- and one wild type or I^+, is like the wild type in producing appreciable amounts of β-galactosidase only in response to induction. Thus strains of constitution $I^+ Z^-/(F)\ I^- Z^+$ or $I^- Z^-/(F)\ I^+ Z^+$ were found both to be inducible like the wild type $I^+ Z^+$. The I^+ gene masked the effect of the I^- gene irrespective of whether it was coupled to Z^+, the gene responsible for the potentiality for β-galactosidase formation. The F′ condition itself does not interfere with the expression of I^-, as shown by the fact that $I^- Z^-/(F)\ I^- Z^+$ behaves like a constitutive mutant. These results are interpreted to mean that I^+ has some positive effect resulting in the *repression* of the activity of Z^+ (and also of Y^+ and the gene A^+ responsible for thiogalactoside transacetylase). Constitutiveness in I^- mutants is regarded as the consequences of the absence of the normal repressor. Substances which are inducers of β-galactosidase formation in the wild type are presumed to act by inhibiting the action of the repressor substance—the product of I^+.

Certain other mutations in *I* have quite a different effect—that of preventing induction of the proteins of the lactose operon under any circumstances. These mutants are called I^s (*s* for *super-repressed*) and

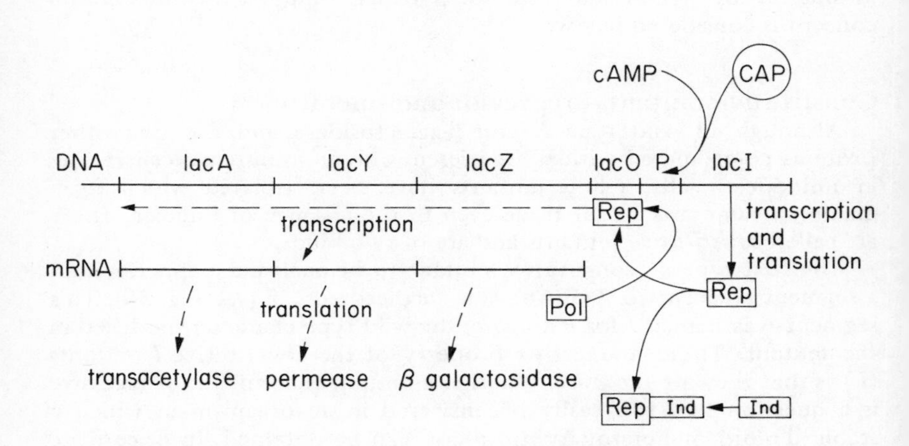

FIGURE 30 The currently favoured version of Jacob and Monod's model for the control of activity of the *lac* operon of *E.coli.* Transcription of the operon into a 3-gene mRNA starts following the binding of RNA polymerase (Pol) to the promoter site P. In the absence of inducer the repressor protein (Rep) binds at the operator segment *lacO* so as to prevent polymerase binding. In the presence of inducer (Ind) a repressor–inducer complex is formed which has no affinity for the operator. A complex of cyclic AMP (cAMP) with cAMP-binding protein (CAP) is in some way necessary for effective polymerase-promoter interaction.

are dominant in diploids also carrying I^+. It was conjectured that I^s mutants produced an altered form of the repressor which no longer had affinity for inducer. As we shall see, this hypothesis has been fully confirmed following the isolation and purification of the protein product of the I^+ gene. The properties and mode of action of this repressor protein, and its altered properties in I^s mutants, are considered below.

The second type of constitutive mutant has very different properties from those of the I^- mutants. For reasons which will become apparent they are called *operator-constitutive* mutants, and are given the symbol O^c. Their most important feature is that the activity of the *lac* genes (Z, Y and A) is only affected by the O^c mutation if they are actually linked to it on the same piece of genetic material. In a partially diploid cell of constitution $O^+ Z^- / (F)\ O^c\ Z^+$ (the so-called *cis* diploid) β-galactosidase is produced constitutively; unlike the I^- mutant in a similar situation the effect of O^c is here dominant over that ofthe corresponding wild type gene. However, in a cell of constitution $O^+\ Z^+/(F)\ O^c\ Z^-$ (the *trans* diploid) the enzyme is formed in appreciable amounts only in response to inducer. Thus O^c alters the action of Z^+ only when in coupling with it, and has no effect on the behaviour of Z^+ on another chromosome in the same cell. There is a precisely analogous effect of O^c on the activity of a linked Y^+ and A^+ gene, producing permease and transacetylase. The O^c mutations are located adjacent to the Z gene, between Z and I (Fig. 30).

TABLE 7

Properties of mutants in the lac *region of* E. coli *in haploid and partially diploid* (*F*-lac) *strains* (adapted from Jacob & Monod, 1961)

Genotype	Formation of β-galactosidase: Presence of inducer	Formation of β-galactosidase: Absence of inducer
$I^+O^+Z^+$ (wild)	+	—
$I^+O^+Z^-$	—	+
$I^+O^+Z^-/(F)I^+O^+Z^+$	+	—
$I^-O^+Z^+$	+	+
$I^-O^+Z^+/(F)I^+O^+Z^-$	+	—
$I^-O^+Z^-/(F)I^+O^+Z^+$	+	—
$I^sO^+Z^+$	—	—
$I^sO^+Z^+/(F)I^+O^+Z^+$	—	—
$I^+O^cZ^+$	+	+
$I^+O^+Z^-/(F)I^+O^cZ^+$	+	+
$I^+O^+Z^+/(F)I^+O^cZ^-$	+	—
$I^sO^+Z^+/(F)I^+O^cZ^+$	+	+

The properties of the various kinds of *lac*-constitutive mutants are summarised in Table 7.

Jacob and Monod proposed a model (Fig. 30) to explain the relationship between the *I* and *O* functions in the control of the *lac* operon as follows. The operon, comprising the *Z, Y* and *A* genes, is transcribed as a unit, starting at the operator end. The repressor protein, the product of the *I* gene, binds to the DNA of the *O* region, so preventing transcription from occurring. The repressor protein is considered to have allosteric properties, responding to certain β-galactoside inducers as allosteric effectors. Paradoxically, lactose as such is not an effector but it can be acted on by β-galactosidase to give rise to a derivative, allolactose, which is. Certain artificial β-galactosides such as the non-utilisable inducer isopropyl-thiogalactoside, or IPTG, do act directly as effectors. On binding to inducer the repressor protein undergoes conversion to a form which no longer binds to the operator. Thus the inducer releases transcription through nullifying the effect of the repressor. This model explained the effects of the various mutations affecting inducibility: I^- mutants no longer make repressor, but I^- is recessive since the repressor can be supplied from another duplicate chromosome fragment; I^s mutants make a repressor which no longer binds to inducer and I^s is dominant since the super-repressor can bind to any *lac* operator in the same cell; O^c mutants have an altered operator no longer binding to repressor and so immune to its action, but the effect only extends to the chromosome actually carrying the mutation.

This model has been confirmed in the most direct and convincing way. As mentioned above, the *lac* repressor can be purified, most easily from a mutant strain which very greatly over-produces this protein. It can be shown that purified repressor protein combines with high affinity with a galactoside inducer of β-galactosidase (the artificial inducer isopropylthiogalactoside, or IPTG, is usually used in this experiment) and *also* with *E. coli* DNA of the *lac* operon. Such DNA can be obtained in greatly enriched concentration in preparations from specialised transducing phage (φ80) carrying the *lac* region of the *E. coli* chromosome. Significantly, the protein did *not* bind to the *lac* DNA in the presence of IPTG. Equally significantly it had poor affinity for *lac* DNA from an operator-constitutive mutant even in the absence of inducer. Furthermore, repressor protein from I^s mutants has been shown to have reduced binding affinity for inducer. Thus the key predictions of the repressor model for operon control were confirmed in cell-free experiments.

Subsequently, it has proved possible to demonstrate that the repressor protein can control transcription of *lac* messenger RNA in a cell-free system. The method depends on detecting *lac* mRNA by its ability to hybridise specifically, by complementary hydrogen bonding, with single-stranded *lac* DNA. The DNA-RNA hybrid duplex can be trapped on filters and, if the RNA is labelled with radioactive atoms, its quantity can readily be measured. DNA of the *lac* operon will, in the presence of RNA polymerase, a mixture of radioactive ribonucleoside triphosphates, cyclic adenosine monophosphate (cAMP) and a cAMP-binding protein, promote the synthesis of radioactive *lac*-specific mRNA. The synthesis is stopped by the addition to the mixture of repressor protein but this inhibition is in turn negated by the simultaneous presence of a suitable inducer such as IPTG. The significance of the cAMP and its binding protein is outlined in the next section.

The promoter region and initiation of transcription

Yet another element in the functioning of the *lac* operon came to light with the discovery of mutants with greatly reduced expression of all the genes of the operon but with normal regulation of the gene activity (such as it was) by inducer. These mutations were mapped by deletion analysis and shown to occupy a short segment (the *promoter*) between the *Z* gene and the operator. They affected only the chromosome in which they occurred and had no effect on the expression of *lac* genes on a separate chromosome fragment. It was inferred that the DNA segment in which these mutations fell acted as an initiation site for transcription, probably by serving as a binding site for RNA polymerase. This hypothesis is confirmed by the demonstration that DNA from a strain with a mutation in the promoter region acts as a poor template for the transcription of *lac*-specific messenger RNA in a cell-free system.

During the last few years many papers have been published on the function of the promoter region and a complex picture is emerging, only the outlines of which can be mentioned here. It seems that the initiation of transcription of the *lac* operon depends on adenosine 2′, 3′ cyclic monophosphate (cyclic AMP or *c*AMP) in association with a specific protein for which the *c*AMP has high affinity. The *c*AMP-protein complex binds to the DNA at or adjacent to the promoter region and, in so doing, probably facilitates in some way the binding of RNA polymerase. In this way, the expression of the *lac* operon is brought under the positive control of yet another small molecule—*c*AMP—the intracellular concentration of which is negatively correlated with the level of carbon nutrition of the cell. This appears to be the basis of the so-called glucose effect, otherwise known as *catabolite repression,* whereby the synthesis of the *lac* proteins, as well as of enzymes for utilisation of other exotic carbon sources, is repressed when more commonly available sugars are in plentiful supply.

Nucleotide sequence of the operator-promoter region

Very recently (1973) great progress has been made in W. Gilbert's laboratory at Harvard University on the direct chemical determination of the nucleotide sequence of the DNA of the *lac* operator-promoter region and of the RNA transcribed from it. This work was made possible by the availability, on the one hand, of DNA containing the *lac* operon isolated from a specialised *lac*-transducing derivative of lambda or ϕ80 phage and, on the other hand, of purified *lac* repressor protein. If repressor protein is added to the DNA and then DNAase added to the mixture, all the DNA is digested to small soluble fragments except the *lac* operator region, which is protected by its tight binding to the repressor. As a complementary approach, DNA from the transducing phage was broken into small fragments by sonic vibration, and those fragments capable of binding to repressor were isolated by using the fact that the DNA-repressor complex will bind to a nitrocellulose filter.

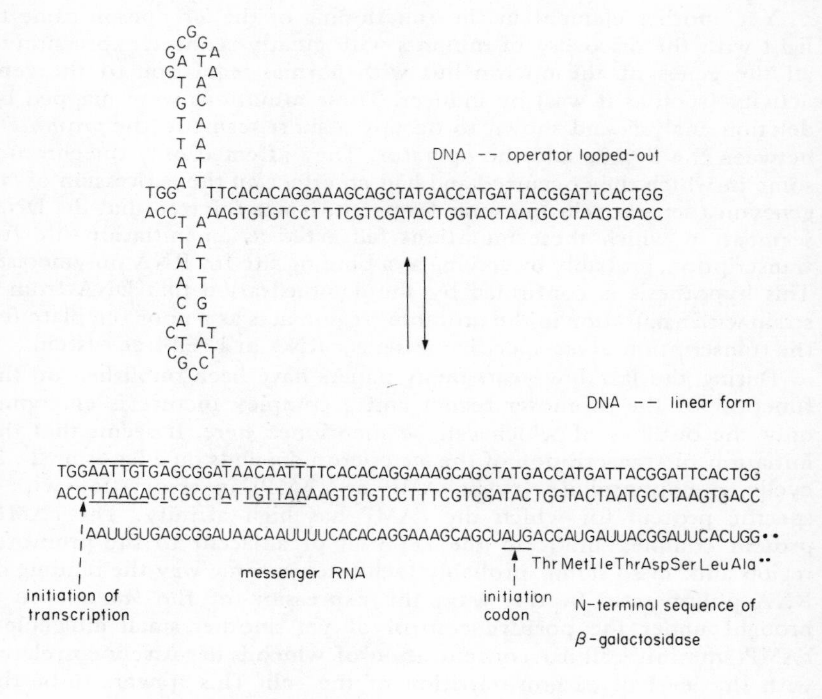

FIGURE 31 The nucleotide sequence of the operator–promoter region of the *E.coli lac* operon, extending to the start of the *lacZ* gene, coding for β-galactosidase. The operator region is drawn in both its linear and its hypothetical looped-out form, made possible by the rotational symmetry of the sequence of the nucleotide pairs. Whether this form actually plays any role in repressor binding is controversial, but the partial symmetry (shown by horizontal bars in the linear form) is in any case probably significant for interaction with repressor. The sequence of the mRNA shown was determined after transcription from the DNA of a mutant strain with an altered promoter which did not require cyclic AMP or cAMP binding protein; it has not been proved that transcription starts at the same place in wild type bacteria. After the work of Gilbert, Maxam and Maizels (ref. p. 142).

The first approach leads to the isolation only of the operator segment and the second to the isolation of the operator included within longer stretches of DNA. The operator segment itself (in the sense of that segment protected from digestion by repressor) turned out to be only 24 base-pairs long—short enough for its base-pair sequence to be determined by chemical methods. The longer segments were used as templates for the transcription of mRNA in a cell-free system. The sequences of these transcripts were determined and were shown to be a family of sequences all starting at the same point near the end of the operator (presumably the promoter, or adjacent to it) but extending for different distances away from the operator. The longer transcripts overlapped into the beginning of a sequence which was evidently the message coding for β-galactosidase, the terminal amino-acid sequence of which is known.

The most notable feature of the operator segment was its high degree of rotational symmetry—in a 21 base-pair sequence the left-hand 10 base-pairs, with two exceptions, were the same as the right-hand 10 pairs rotated through 180 degrees. With such a sequence the operator DNA could in principle adopt a secondarily base-paired, looped-out structure as shown in Fig. 31, and it is suggested that it is to this structure that the repressor protein specifically binds. This conjecture is reinforced by the finding that several operator mutants, with reduced affinity for repressor, have single base-pair changes in the operator region which disrupt the symmetry and presumably reduce the stability of the looped structure.

The results, which are summarised in Fig. 31, represent a very exciting development and an impressive advance towards the full understanding of the control of at least one operon in molecular terms.
5-phosphate (a normal intermediate in carbohydrate metabolism) and a closely linked gene *C* controlling a regulator protein. Furthermore, some mutations in *C* result in *constitutive* formation of the three

OTHER INDUCIBLE ENZYME SYSTEMS—THE *ARA* OPERON

Although the *lac* operon has been more extensively studied than any other and provides the most influential model or paradigm for the regulation of gene action, several other inducible systems involved in utilisation of special carbon or nitrogen sources are known. Some of these have features not found in the *lac* operon and cast some doubt on the adequacy of Fig. 30 as a general model. In this connection the operon concerned with utilisation of the pentose sugar arabinose must be mentioned.

Here the situation is similar in some respects to that described for *lac*: there are three adjacent genes specifying the three arabinose-induced enzymes needed for the conversion of L-arabinose to xylulose-

enzymes in the absence of the inducer arabinose. Many ingenious experiments, for details of which the reader must refer to the publications of Engelsberg and his colleagues (see Vogel, ref. p. 142), have shown, however, that the *C* gene product not only functions as a repressor in the absence of arabinose but is also positively necessary for induction in the presence of arabinose. Mutations involving the substantial deletion of *C* cause loss of the repressor function (as shown by the recessiveness of such mutations with respect to this function in partial diploids) *and* fail to show induction of the three enzymes by arabinose. A further mutation was selected in these *C*-deletion strains to restore activity of the *ara* genes; this turned out to map at one end of, and just outside, the block of three genes and had the effect of allowing transcription of the operon independently of arabinose and the *C* gene product. The segment within which this constitutive mutation occurred is regarded as a site for initiation of transcription, and the mutant is called I^c, for initiator-constitutive. I^c does not render transcription immune to the repressor action of the C^+ product, although certain deletion mutations which cut into an 'operator' segment immediately adjacent to *I* on the side away from the structural genes do so (Fig. 32). Thus it is believed that the C^+ product is a protein with dual action, *repressing* transcription in the absence of arabinose by combining with an operator segment, and *activating* transcription in its arabinose-bound form by combining with an adjacent initiator segment. A curious difference from the *lac* system (Fig. 30) is that the presumed site of polymerase attachment *I* is *between* the structural genes and the operator instead of being separated from them by the operator. As in

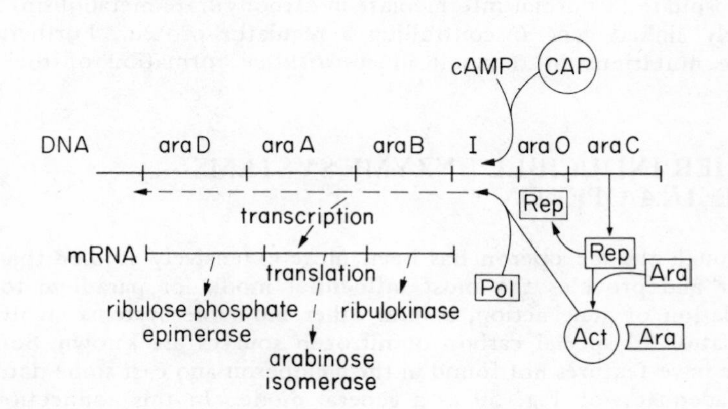

FIGURE 32 The system of regulation of the *arabinose* operon in *E.coli.* Symbols as in Fig. 31 except Act = activator, Ara = arabinose. Based on Engelsberg in *Metabolic Regulation* (see bibliography, p. 142).

the case of the *lac* operon, cyclic AMP and a cAMP binding protein are also needed for transcription.

The dual role (positive as well as negative) of the regulator protein is the main feature of the *ara* system not seen in the *lac* system; it is too soon to say which situation is the more general or whether yet other systems of control await discovery.

THE RELATION BETWEEN REPRESSIBLE AND INDUCIBLE SYSTEMS

Although, at the beginning of this chapter, enzyme induction and repression were mentioned as two distinct phenomena, the interpretation of induction of the activity of the *lac* operon as a *release of repression* suggests that the difference between the two types of control may consist merely in the way in which an allosteric repressor protein responds to the binding of the small molecular effector. In the *lac* system, as we have seen, binding of inducer renders the repressor protein ineffective as a repressor while, in the *ara* case, it converts the repressor protein to an essential initiation factor. Systems of enzymes involved in synthesis, whose formation is repressed by their respective end products, seem also to be under the control of repressor proteins, the difference being that the binding of the effector molecule (the end-product or a derivative of it) promotes the repressor action rather than neutralising it.

It follows from this that mutants in both kinds of system which result in loss of the repressor protein will result in *unconditional* enzyme formation—constitutive in the one case, non-repressible in the other. Mutants producing altered repressor protein, no longer responding to the effector, will have opposite effects in the two kinds of system. They will be super-repressed non-inducible mutants (like *lac* I^s mutants) in the catabolic inducible systems but non-repressible enzyme producers in the anabolic systems.

Several operons concerned with biosynthetic pathways could be used to illustrate the parallels with, and differences from, the 'standard' operon model derived from the *lac* system. Most of the good examples concern amino-acid synthesis, especially histidine, tryptophan (see p. 40) and arginine. We shall concentrate on the histidine system of *Salmonella typhimurium*.

THE HISTIDINE OPERON OF *SALMONELLA TYPHIMURIUM*

In the last chapter mention was made of the remarkable studies of Demerec, Ames, Hartman and their collaborators on histidine auxotrophs of Salmonella. Fig. 25 showed the arrangement of the genes controlling the enzymes of histidine biosynthesis. The significance of the close linkage of these genes is the same as that of the genes of the *lac* and *ara* operons, namely that they are transcribed into a single molecule of messenger RNA. This large polygenic messenger has actually been identified as a high molecular weight species, present in cells in which the histidine-synthetic enzymes were derepressed but absent from repressed cells; it is of the size expected if it carries the coding information for the amino-acid sequences of all the histidine-synthesising enzymes.

Co-ordinate regulation

Just as the enzyme products of the *lac* operon are produced in constant ratio, that is co-ordinately, so too are the enzymes of histidine biosynthesis in *Salmonella typhimurium.* Ames and his group showed that all these enzymes were present in only small amounts in wild type cells grown on minimal medium. However, any histidine auxotroph lacking just one enzyme produced large amounts of all the others when starved of histidine. This was taken to mean that the wild type normally produces a high enough concentration of histidine within the

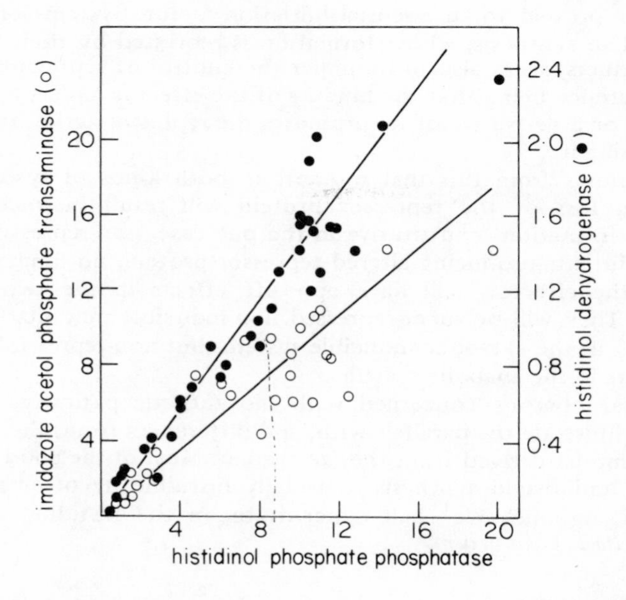

FIGURE 33. Co-ordination of production of enzymes of histidine synthesis in Salmonella (redrawn from Ames and Garry, 1959). Mutants of class E (cf. Fig. 25, p. 76) were grown on various levels of formylhistidine to give various degrees of repression of the enzymes. Each point represents an enzyme measurement in a separate culture. Ordinates show enzyme activities, in units which are different for each enzyme. Despite some scatter of the points, the transaminase and the dehydrogenase both show an obvious tendency to be produced in constant ratio to the phosphatase, and hence in constant ratio to each other.

cell to repress, to a large extent, the formation of histidine synthesising enzymes. In other words, *S. typhimurium* has a greater potential for synthesising histidine than it actually needs when growing in the standard synthetic medium, and this potential is held in check by partial repression actuated by intracellular histidine. By varying the histidine supply different degrees of repression can be obtained. Ames found that all the enzymes (or at least those five of them which could easily be measured) could vary greatly in absolute amount from one culture to another but that their *relative* amounts remained very nearly constant (Fig. 33). This was the first example of co-ordinate repression and de-repression of a group of enzymes coded by a block of adjacent genes, and it found a ready explanation in terms of control of the synthesis of a polygenic (or polycistronic) messenger.

The nature of the repressor

To what extent are the other essential elements of the *lac* operon control system—the specific repressor protein and the operator segment to which it binds—to be found in the *his* system? So far as the repressor is concerned the situation is complicated. An embarrassingly large number of genes have been found to be capable of mutating to cause non-repressible synthesis of the histidine enzymes. Such mutants are readily selected as resistant to the histidine analogue thiazole-alanine, which is not an acceptable substitute for histidine but is still sufficiently like it to cause repression of the histidine synthetic enzymes in the normal bacterium. It is still not certain which class of thiazole-alanine resistant mutants, if any, codes for the hypothetical repressor protein. One class appears to be mutant in the structure-determining gene for histidyl-tRNA synthetase, while another appears to represent a gene for histidine-specific tRNA, since mutations in it are shown to cause changes in the base sequence of this tRNA. From these mutants it is deduced that histidyl-tRNA, rather than histidine itself, is probably the *co-repressor* which binds to and so actuates the repressor protein; in this event any mutation altering histidine-specific tRNA or the formation of its histidyl derivative might be expected to affect repression. The most intriguing recent development is a report that histidyl-tRNA forms a specific complex with the first enzyme of the pathway, phosphoribosyl-ATP synthetase. Furthermore, it seems that certain mutations desensitising this enzyme to end-product inhibition affect the timing, if not the eventual extent, of derepression of the operon. Though the picture is still far from clear at the time of writing, there is quite a strong suggestion that the PR-ATP synthetase (known to be an allosteric enzyme) is involved in the repression mechanism. The possibility that enzymic proteins may be involved in regulation of their own synthesis and that of other enzymes of the same pathway is one for which there is evidence in several other cases, particularly in fungi. Such a dual role of proteins, as enzymes and as regulators of protein synthesis, may be of key importance in the complex circuitry of cell metabolism.

The operator

Among the thiazole-alanine resistant mutants which display non-repressible formation of the enzymes of histidine synthesis is one class which has the crucial property expected of operator mutants, conferring constitutive activity on the *his* genes linked to the mutation on the same chromosome (i.e. in the *cis* position) but not on *his* genes carried on a separate chromosome fragment (i.e. in *trans*). The other thiazole-alanine resistant mutant classes (mentioned above) are recessive both in *cis* and in *trans*. The presumptive operator mutants have been mapped at one end of the operon, beyond but apparently adjacent to the *hisG* gene (Fig. 25).

It has also been shown that several kinds of mutant in which one end of the operon has been deleted fail to form any of the histidine enzymes. The deletions are of various lengths, but they all include a section of the *hisG* gene and overlap into the supposed operator region. In the light of the discovery of the promoter region in the *lac* operon it seems likely that these pleiotropic enzyme-negative deletion mutants have lost a promoter as well as the operator region. In an illuminating extension of these observations, Ames, Hartman and Jacob selected revertants from one of the pleiotropic deletion mutants (*his 203*) which had regained the activity of the *hisD* gene, permitting them to grow on histidinol instead of histidine (cf. Fig. 25, p. 76). A number of such revertants were found, and they all turned out to have two additional unlooked-for properties. Firstly, they had all regained activity not only of *hisD* but of all the other genes of the operon with the exception of *hisG* (which had been irreparably damaged by the original deletion). Secondly, the enzymes controlled by the operon were produced constitutively and could no longer be repressed by histidine.

Genetic analysis showed that these non-repressible revertants were of two distinct types. In both, the *hisG* gene had been effectively deleted. In the first the original *his 203* deletion had been extended to cover the greater part of *G* and, presumably, an additional segment to the right of the operon. The second was an unstable type in which it was shown that the original *his 203* genome was still present, with an *additional* fragment of genetic material carrying the *his* operon (minus *G*) with restored activity. This remarkable kind of fragment, capable of maintaining itself by replication, is an example of a *plasmid*; more examples will be seen in the final chapter. Presumably the starting strain already harboured a cryptic plasmid to which the *his* operon could be transferred by a rare chance event.

The two kinds of revertant are represented diagrammatically in Fig. 34. What they have in common is that the *his* operon is in each case attached at its right end to material with which it is not normally associated. It seems that in each case the surviving *his* genes have been placed under the control of *another* (unknown) promoter which functions constitutively under all normal conditions of culture.

It is worth mentioning in passing that recent experiments of Beckwith and his colleagues have demonstrated transposition of the *lac*

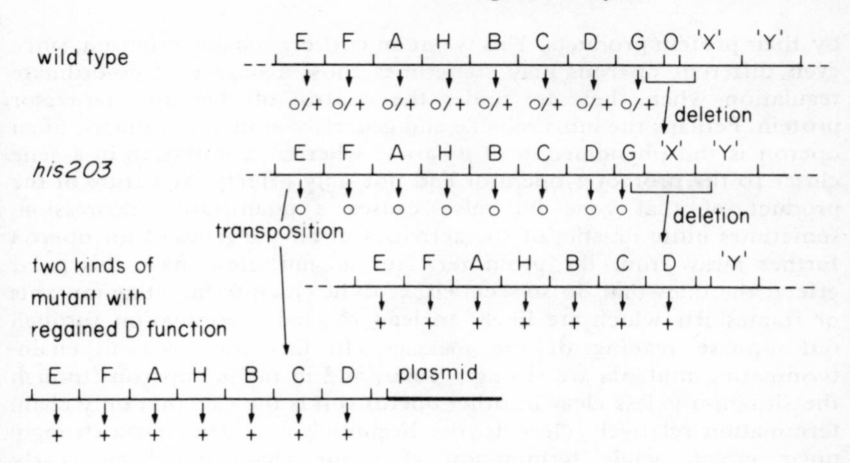

FIGURE 34 Structural alterations in the *his* region of the Salmonella genome which modify the control of the activity of the *his* operon. In the wild type the operon is active or not depending on the histidine supply. In *his*203 the genes are permanently 'switched off', supposedly because the 203 deletion has eliminated the *his* operator (*o*) and joined the operon to a neighbouring operon ('x') which is inactive under all known conditions. Either further deletion or transposition of the *his* segment to a plasmid can join the operon to another ('y') which is *active* under all known conditions. The G gene is not active in either kind of secondary mutant, and it is presumed to have been deleted. After Ames, Hartman and Jacob (1963).

operon so as to fuse it to the operon concerned with tryptophan biosynthesis. In these transposition mutants the *lac* genes are under the control of the *trp* promoter and operator, and display the bizarre property of being repressed by tryptophan along with the *trp* genes.

POLARITY WITHIN OPERONS

Not all clusters of genes of related or sequential function are operons in the sense of being units of transcription into single mRNA molecules. Examples are known where such a cluster comprises two separately transcribed operons. What, then, are the criteria for distinguishing truly single operons from multiple-operon clusters?

One criterion has already been mentioned—that of the occurrence of operator or promoter mutations affecting the expression of the entire cluster of genes if they are members of a single operon. Operator mutants are not always easy to obtain, however, and promoter mutants tend to be even more elusive. A second property is the co-ordinate regulation of the genes of an operon, seen in the constant ratios shown by their protein products. This is not an entirely reliable criterion, since even different operons may sometimes show a degree of co-ordinate regulation when they are under the control of the same repressor protein. Perhaps the most reliable and generally available hallmark of an operon is the phenomenon of *polarity*, whereby a mutation in a gene closer to the promoter/operator end not only affects the nature of the product of that gene but also causes a quantitative depression, sometimes quite drastic, of the activities of all the genes of the operon further away from the promoter. Not all mutations have this polar effect; the ones that do so seem either to be chain-terminating mutants or frameshifts which are likely to lead to chain termination through out-of-phase reading of the message. In fact not even all chain-terminating mutants are strongly polar, and in the *lac* operon (though the situation is less clear in other operons) it is the rule that only chain termination relatively close to the beginning of a gene has a strongly polar effect, while termination after the chain has been nearly completed has only a weak polar effect, if any.

The mechanism of polarity has been the subject of much controversy. At present the most likely explanation seems to be that the mRNA beyond a premature termination codon is exposed to attack by ribonucleases because not protected by ribosomes. In bacteria it seems that ribosomes attach in a densely packed train to the mRNA, even while the latter is peeling off the DNA template, so that the only sections of the messenger not covered by ribosomes will be those where there is a substantial gap between a termination codon and the next ribosome attachment site (usually at the beginning of the next gene). Where such a gap occurs, the hypothesis states, nuclease molecules can gain an entry and degrade the messenger all the way back to its origin in the DNA as fast as it is formed. Thus ribosomes will have a much reduced chance of attaching at the translation-initiating codons of any gene transcribed after the polar mutation.

Whether this turns out to be a complete account of the mechanism or not, the occurrence of a polar effect of a chain-terminating mutation can only mean that the genes on which the polar-effect is exerted are part of the same translational, and hence transcriptional, unit as the gene in which the mutation occurred.

THE GENERAL SIGNIFICANCE OF OPERONS AND OF CLOSE LINKAGE

Operon organisation of functionally related genes is common in bacteria, but there are also numerous examples of genes involved in the same metabolic pathway occurring in scattered locations. In these latter cases the scattered, metabolically related genes may still be subject to regulation by a common repressor protein and they are then sometimes referred to as constituting a *regulon* rather than an operon. The genes controlling arginine synthesis in *E. coli* are a case in point. It is easy to see in a general way that a relatively loose and flexible co-ordination may often be more appropriate than a tightly geared one, since some of the gene products may have at least occasional functions not shared by all the others. Only when a number of enzymes function exclusively in series and never individually is an operon type of organisation likely to be established by natural selection. We might expect this in the case of unbranched and highly specialised metabolic pathways; the histidine and tryptophan pathways are good examples.

Even when metabolically related genes do not constitute a single operon it is quite common to find them closely linked. In particular it seems to be rather common, though not invariably the case, that the gene specifying the repressor protein for a particular operon is closely linked to that operon even though it could function just as well at a distance or even, as experiments on partial diploids have shown, on a separate chromosomal fragment. One tends to assume that the close linkage here has some significance, but it seems more likely that it reflects evolutionary processes rather than the optimal function of the contemporary organism. To take the *lac* operon as an example, this is a set of genes of rather specialised use, possessed by *E. coli* but absent from some other related bacteria, for example *Salmonella typhimurium*. No doubt it has, in the course of bacterial evolution, often been transferred by natural conjugation or transduction from one cell type to another. The repressor gene *lacI* would confer no selective advantage if transferred separately from the genes which provide its sole *raison d'être*. One would expect there to be powerful selection in favour of the establishment and maintenance of close linkage between regulator genes and the operons they control, so that the whole apparatus can be transferred as a self-contained unit. The same kind of explanation probably applies to the *ara* operon and its controller gene *araC* as well as to numerous other cases of close linkage between functionally related but separate operons.

In eukaryotic organisms the importance of operons of the bacterial kind, and even whether they occur at all, is still in doubt. There are, especially in fungi, numerous examples of clusters of functionally related genes showing some of the features of operons, notably the occurrence of pleiotropic polar mutants. In no case, however, is the interpretation completely unequivocal, though it could become so in the next few years. For a discussion of this question the interested reader is referred to the text on *Fungal Genetics*, or the article by G.R. Fink in *Metabolic Regulation*, cited in the Bibliography, p. 142.

REGULATION OF TRANSCRIPTION DURING DEVELOPMENT

In the preceding discussion we have encountered examples of both negative and positive controls of transcription acting in response to changes in nutrients. The negative controls are exercised by repressor proteins, combining with DNA at operator sites and in some way blocking transcription. The nature of the positive controls is still not understood, but one possibility is that specific sigma factors (cf. p. 82) are involved, combining with RNA polymerase so as to activate it for attachment to specific transcription-initiating sites (promoters). Such positive control proteins might in their turn be subject to control by allosteric effectors. The effectors, and for that matter the control proteins are, of course, themselves the products of the activity of genes, and it is easy to imagine a 'cascade' of effects, with the activity of one gene turning on (or repressing) the transcription of a second, the activity of which affects the transcription of a third, and so on. Such complex series of effects could form loops, resulting in either positive or negative feedback, and such circuits could very well be involved in the establishment and stabilisation of differentiated states of cells in general. By far the best worked-out model is provided by the very detailed studies of bacteriophage *lambda* (cf. pp. 42-44), where just such self-stabilising or self-reinforcing circuits are involved, on the one hand, in the maintenance of the stable lysogenic state and, on the other, in the rapid yet orderly sequence of gene activities involved in prophage release and multiplication, phage maturation and cell lysis. For details of this exceedingly complex but beautiful system, the reader is referred to the multi-author monograph edited by Beckwith and Zipser and the review by Herskowitz, cited in the Bibliography, p. 142.

7 Plasmids

One of the most intriguing findings to have emerged from the study of bacterial genetics is the widespread occurrence in bacteria of *extra* pieces of DNA, separate from the main chromosome and replicating independently of it. These *plasmids*, as they have come to be called, are of the nature of 'optional extras' so far as the bacterium is concerned; the genes which they carry are not essential for the life and growth of the cell but they do confer additional properties which may in some circumstances assume crucial importance.

Many aspects of the natural history of plasmids can be illustrated by reference to the F-factor of *E. coli* (cf. pp. 37-38), although there are some respects in which it is atypical. In the remainder of this chapter we will review in turn the main characteristics common to all plasmids and the great variety of properties in which they differ from each other and by reference to which they can be classified.

INDEPENDENT REPLICATION OF PLASMIDS

The one absolutely essential attribute of a plasmid is the capacity to replicate independently of the main chromosome. To use one terminology, a plasmid must be a *replicon* in its own right and, in particular, possess a *replicator* segment or region in its DNA which serves as a point of initiation for the action of the appropriate DNA polymerase. This is not a property of any piece of DNA; as we saw in Chapter 3, random fragments of *Salmonella typhimurium*, such as are concerned in abortive transduction, have no capacity for independent replication.

The separate replication of plasmid and chromosome can be shown in a number of ways. For example there are a number of circumstances in which the replication of the one can be inhibited without the other. Several compounds capable of intercalating into the stack of base-pairs of DNA, including acridine orange and ethidium bromide, selectively inhibit replication of F without affecting that of the chromosome. Thus these drugs provide a means of ridding a culture of the F-factor. On the other hand, certain temperature-sensitive chromosomal mutations prevent initiation at 42°C of replication of the chromosome while leaving replication of the F factor unimpaired.

Even more interestingly, the replication of one plasmid, *Col* E_1 appears to be dependent on a different DNA polymerase from that used for *E. coli* chromosomal replication. Replication of *Col* E_1 fails to occur in mutant (*polA*) cells deficient in DNA polymerase I (the original Kornberg enzyme), though such mutants are well able to replicate their main chromosome. On the other hand, temperature-sensitive mutants with a heat-labile form of another enzyme, DNA polymerase III, fail to replicate the chromosome at 42°C but can support replication of *Col* E_1 at this temperature.

PHYSICAL FORM OF PLASMID DNA

On p. 43 reference was made to the idea that *lambda* bacteriophage DNA can form a closed loop, and how such a hypothesis helps to explain its integration into the chromosome. There is, in fact, excellent physical evidence that the F factor, together with all other plasmids so far as is known, also has such a form. DNA from cells harbouring an F factor, an F′ factor or some other plasmid can be fractionated to yield a generally uniform population of closed-loop DNA molecules, much smaller than the DNA of the chromosome and readily visible under the electron microscope after the DNA has been coated with basic protein to increase its thickness and 'shadowed' with heavy metal ions to increase its electron density. This microscopic technique (due to Kleinschmidt) reveals that the plasmid DNA is in the form of twisted loops, the DNA duplex being thrown into a higher-order coil (Fig. 35). Only if one strand of the duplex is broken (or 'nicked') by enzyme action are the twisted loops free to untwist to an open, really circular form. One important characteristic of twisted closed-loop DNA is that

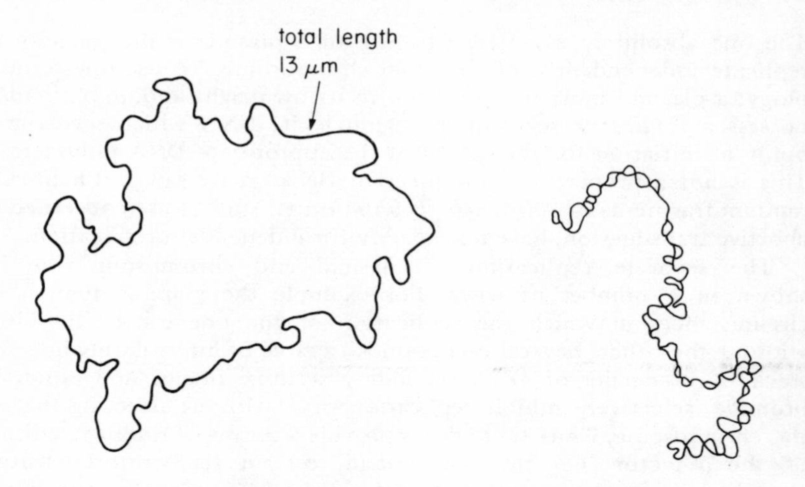

FIGURE 35 (A) The electron microscope appearance of the twisted-circle and open-circle ('nicked') forms of a small *E.coli* resistance-transfer plasmid (drawn from P.Kontomichalou, M.Mitani and R.C.Clowes, *J.Bact.*, **104**, 34-44, 1970).

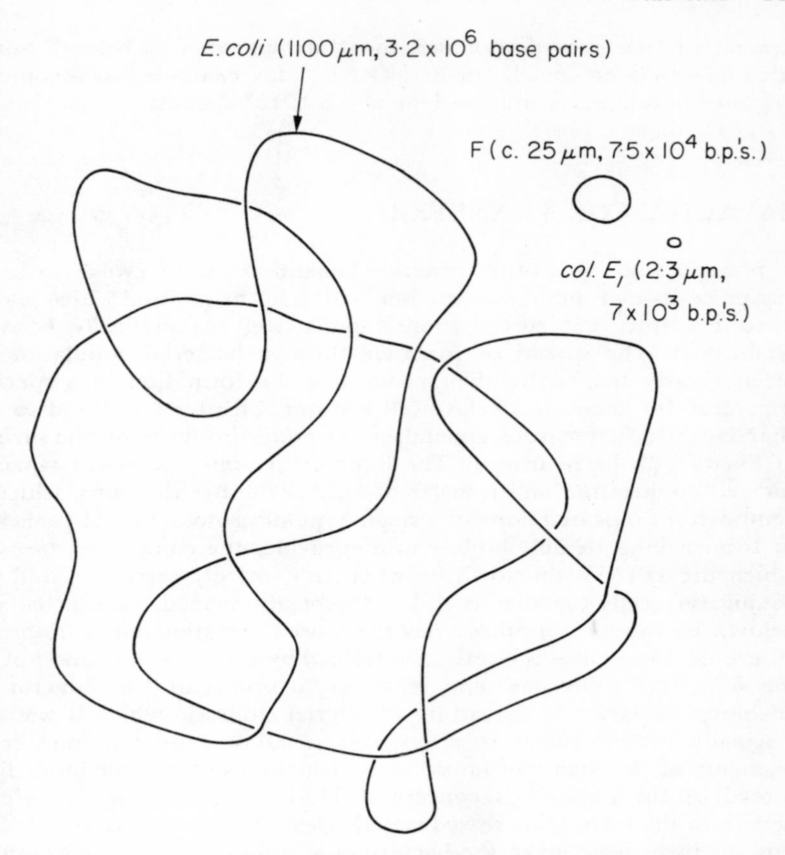

(B) Comparative sizes of the *E.coli* chromosome and two plasmids, one large and one small, all drawn as extended loops of DNA.

it binds ethidium bromide less readily than do the linear fragments of broken-up chromosomal DNA which constitute the bulk of the DNA isolated from bacterial cells. The reason for this is that bound ethidium can only be accommodated through some unwinding of the DNA duplex and this occurs much less easily in the case of a twisted closed loop. The effect of the restricted ethidium binding of plasmid DNA is that this DNA in the presence of ethidium remains more compact and dense than does the bulk of the DNA. This permits the plasmid DNA to be isolated very cleanly by density-gradient centrifugation in the presence of ethidium bromide.

F-factor DNA turns out to consist of a closed loop of contour length about 25 μm and molecular weight approximately 5×10^7, enough for something of the order of 50 to 100 genes. It is not at all clear how one can account for all this DNA in terms of encoded information, though a number of functions determined by it are known (see below). Some other plasmids are much smaller; *Col* E_1, for example has a contour length of only 2.3 μm and a weight of 4.5×10^6 daltons.

CAPACITY FOR TRANSFER

A plasmid, like any other genetic element, tends to evolve so as to maximise its own numbers; any innovation in the plasmid which allows it to multiply and spread more rapidly will automatically become established. The spread of plasmids through bacterial populations is often greatly helped by their promoting the formation of a specific apparatus for their own cell-to-cell transfer. Chapter 3 referred to the characteristic filamentous appendage, or *pilus*, produced on the surface of *E. coli* cells harbouring F. The F-pili or 'sex-pili' confer the capacity for cell conjugation and transfer of DNA. Whether the pilus, which is composed of repeated units of a single type of protein, helically stacked to form a long flexible hollow tube, provides the connection *through* which the DNA is threaded, or whether it merely serves to hold the conjugating cells together is not yet generally agreed. As we shall see below, the former hypothesis has the more interesting implications. At all events, the F-pilus is clearly determined by the F-plasmid and not by the *E. coli* chromosome, and it serves to propagate the F-factor by enabling the latter to spread into bacterial clones in which it was not originally present. That it serves also to allow a limited transfer of segments of the main chromosome is incidental so far as the immediate spread of the plasmid is concerned, although it may well confer a benefit in the form of increased genetic flexibility on the bacterium and this, in turn, may make the bacterium a more expansive environment for the plasmid. Pilus formation is not the only function determined by F necessary for cell-to-cell transfer of the plasmid; several other genes carried in the plasmid and necessary for transfer have been identified through studies of non-transmissible mutants of the plasmid, but the precise functions of these other genes are less well understood.

Comparative studies of a variety of plasmids have revealed that many of them carry the genetic information for the formation of a pilus of the F type. Others, however, exemplified by the *Col I* plasmid, produce pili of a different type. The two types may be distinguished in two ways. First, antibodies directed against F-pili will inhibit F-mediated genetic transfer in *E. coli* while antibodies directed against *Col I*-type pili will not. Secondly, each type of pilus provides specific sites of attachment for a variety of tiny bacteriophages, some filamentous in form with single-stranded DNA as genetic material and some spherical with single-stranded RNA as genetic material. Thus F^+ and Hfr cells of *E. coli* are susceptible to infection by F-specific phages. The set of phages which will attach to F-type pili are different from the ones

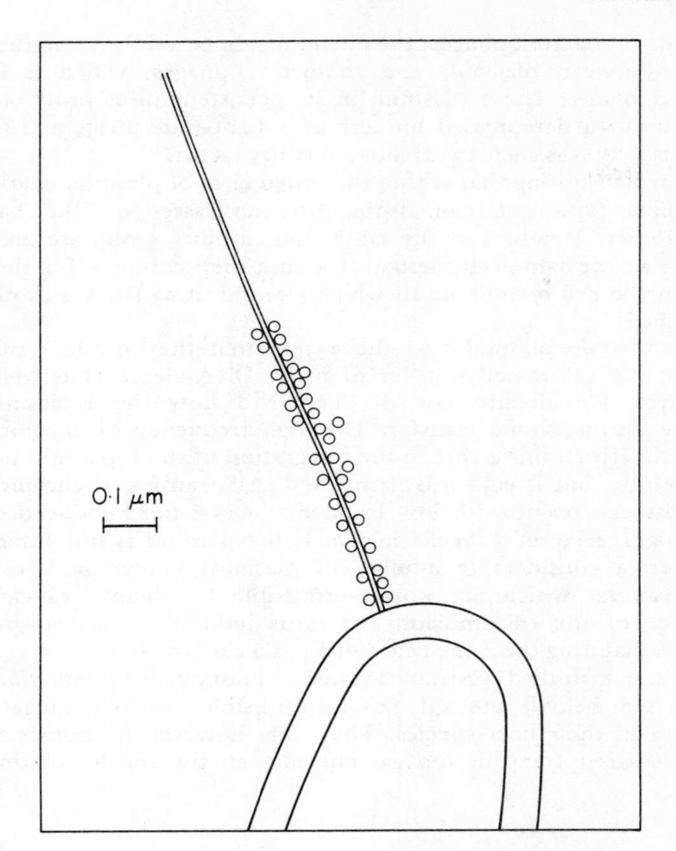

FIGURE 36 Electron microscope appearance of a F-pilus to which both male-specific spherical RNA phage (MS2) and filamentous DNA phage (f1) is attached. The filamentous phage is attached end-on at the tip of the pilus and is distinguishable from it only through not being coated by MS2 particles. Drawn from the photograph in Hayes (*Genetics of Bacteria and their Viruses*) originally provided by Drs. L.G. Caro and D.P. Allison.

which attach to *Col I*-type pili, and the two types of pili can be distinguished on this basis (Fig. 36).

One respect in which the F-plasmid is not typical of transmissible plasmids in general is in its persistent high-level pilus formation. Other transmissible plasmids, including some others of the F-pilus type, seem, after they have inhabited a cell for a short time, to bring about a repression of their own pilus production. In this property they are reminiscent of temperate bacteriophages such as *lambda* which determine the formation of a repressor of their own vegetative functions. The significance of self-repression in the natural history of plasmids may well be that it makes the host cell less vulnerable to infection by pilus-specific bacteriophages; the phenomenon certainly strengthens the analogy between plasmids and temperate phages, which is further discussed below. The F-plasmid, in its persistent pilus production, is analogous to a derepressed mutant of a temperate phage and this, of course, is why it is such an effective fertility factor.

It is worth noting that within the broad class of plasmids sharing the same pilus type one can distinguish sub-classes on the basis of *compatibility*. Members of the same compatibility group are unable to replicate in the same cell, perhaps because they compete for the same niche on the cell membrane to which all replicating DNA is thought to be attached.

Transmissible plasmids, to the extent that they produce pili, can promote the cell-to-cell transfer of other DNA elements as well as of themselves. We already saw in Chapter 3 how the F-plasmid can promote chromosomal transfer. The high frequency of chromosomal transfer in Hfr strains is due to the integration of the F-plasmid into the chromosome, but it is at least suspected that transfer of chromosomal fragments can occur with low frequency across the F-induced cell-to-cell bridges even in F^+ cells in which the plasmid is not integrated. Moreover, a considerable number of plasmids known in *E. coli* and allied bacteria which are non-transmissible by themselves can take advantage of the transmission apparatus induced by a transmissible plasmid inhabiting the same cell. *Col* E_1 is a case in point.

The much-studied resistance-transfer plasmids of *Staphylococcus aureus* (see below) are all non-transmissible, since conjugation is unknown in their host species. They can, however, be transferred by phage-mediated transduction as, indeed, can the smaller plasmids of *E. coli.*

INTEGRATION INTO THE BACTERIAL CHROMOSOME

One of the most distinctive features of the F-plasmid is its ability to integrate into the bacterial chromosome at many different sites. The mechanism involved is the same as that illustrated in Fig. 18 (p. 43) for *lambda* bacteriophage. This property explains the origin of high-fertility (Hfr) strains of *E. coli,* since the integrated F-factor can mobilise the whole chromosome, or a substantial segment of it, for transfer to the F^- cell.

The ability to integrate reversibly into the chromosome is the defining characteristic of an *episome,* and at one time virtually all plasmids were supposed to be episomes. It now seems, however, that episomal properties, though so characteristic of F and of many temperate bacteriophages such as *lambda*, may not be widely shared by plasmids in general. There is very little evidence for the chromosomal integration of the various colicinogenic and resistance-transfer factors to be mentioned below. What the F-plasmid possesses that makes it more prone to integration than other plasmids, especially with such an apparent lack of specificity as to site, is still something of a mystery.

One very important consequence of chromosomal integration of F, and one which is probably of great significance for the evolution of plasmids in general, is the occasional origin of F′ factors (see pp. 75, 95) in which, through inexact excision, the original plasmid is augmented, or partly replaced by, a block of genes from the bacterial chromosome. The analogy with the origin of the defective transducing *lambda* phage (p. 43) is a close one. This property both of F and of *lambda* provides a plausible model for the acquisition of new genes by plasmids of any kind capable of even rare chromosomal integration.

ADDITIONAL GENE FUNCTIONS OF PLASMIDS–COLICINS AND RESISTANCE

Since the discovery of the F-plasmid, which carries information for only one well characterised product–the F-pilus protein–numerous other transmissible plasmids with a much wider repertoire of gene activities have been described. They fall into two not necessarily mutually exclusive categories, depending on whether they are detected by their determining the formation of *bacteriocins* or by their conferment of *resistance* to various antibacterial agents. Bacteriocins are proteins produced by certain bacteria which, in one way or another, kill other bacteria. The bacteriocins which are definitely known to be determined by plasmids are all produced by *E. coli* or related enteric bacteria, and are known as *colicins.* It was recognised in the 1950's, mainly as a result of the pioneering work of the Belgian microbiologist P. Fredericq, that the capacity to produce colicins was conferred by *colicinogenic factors,* which could be transmitted in infective fashion from cell to cell. In more recent years G.G. and E. Meynell have shown that many colicinogenic factors resemble the F-factor in determining the production of pili; these are sometimes of the same type as F-pili, but certain colicinogenic plasmids, of which *Col I* was the prototype, are associated with a second, distinct type of pilus. The distinction between the two types has already been touched upon (p. 114). Whichever type of pilus is produced, it enables cell conjugation to take place and the plasmid to pass from one cell to another. There are, however, two respects in which most colicinogenic plasmids differ from the F-factor. Firstly, as already mentioned, they produce pili much less freely than does F because they tend to repress their own activity in this regard, rather as a temperate bacteriophage produces a repressor of its own unrestricted multiplication. The F plasmid is unusual in being apparently permanently derepressed and hence an unusually efficient promoter of conjugation. The second way in which the best known colicinogenic factors differ from F is in the lack of evidence for their ability to integrate into the bacterial chromosome and hence to act as Hfr agents.

Most of what has just been said about colicinogenic factors applies equally to *resistance transfer factors*. These were first discovered by Watanabe in Japan in the early 1960's, but they have since been shown to be extremely widespread, especially in bacteria isolated from hospital patients. Many, though by no means all, are associated with the formation of pili, of F- or I-type in different cases,[1] and are thus transmissible by cell conjugation. Their common characteristic is that they confer on their host cells resistance to drugs, such as penicillin, chloramphenicol, tetracycline or streptomycin, or to toxic metal ions such as mercury or cadmium. It is frequently found that a resistance transfer factor carries determinants for resistance to several of these antibacterial substances at once. Indeed they may sometimes also confer resistance to bacteriophages of certain kinds and even be colicinogenic into the bargain. For their part, colicinogenic factors must normally confer resistance to the colicins which they produce—otherwise they would soon perish for want of a living cell to inhabit. It thus becomes apparent that some plasmids carry many different genes, some of which can be extremely advantageous, under certain circumstances, to the bacterial cells carrying them. We saw above that there is enough DNA in an F-factor for about 50 to 100 genes; many resistance-conferring plasmids are somewhat smaller than this but still contain enough DNA for the various functions they are known to determine and more besides.

The spread under conditions of modern multiple chemotherapy of plasmids conferring multiple resistance to antibiotics has become a serious problem. Even the non-transmissible (non-pilus-producing) resistance-conferring plasmids (of which many are known) sometimes spread from one bacterial strain to another by taking advantage of the presence of another transmissible plasmid, or by transduction. Generalised transducing phages have a very wide distribution among bacterial species and it seems likely that transduction is of importance in spreading the plasmids found in the important pathogen *Staphylococcus aureus*. Many of the staphylococcal plasmids confer resistance to cadmium ions, and they may, in addition, be associated with resistance to penicillin or erythromycin or some other antibiotic, or several of these at once. Their extrachromosomal nature is shown by the fact that their loss can be induced by certain conditions which do not affect chromosomal replication or stability. For example growth at high temperature (44°C) leads to loss of plasmids from a high proportion of *S. aureus* cells, resulting in the simultaneous loss of all the resistance characters without impairment of any of the functions essential for normal growth.

[1] Yet other classes of pili have recently been described; one of the most studied is called type N.

ORIGIN AND EVOLUTION OF PLASMIDS

There are numerous indications that plasmids often, if not always, descend from genetic material foreign to the bacterial cells in which they are found, and there is a good case to be made for regarding them as specialised or degenerate viruses.

There are many indications of the exotic origins of plasmids. We have already mentioned some of the evidence that plasmid replication may in some cases have very different requirements from those for chromosomal replication. Again, it is sometimes found that the plasmid DNA has a base composition, and hence a buoyant density, quite distinct from that of the chromosomal DNA. Moreover, the genes carried by plasmids may have activities which are quite foreign to the host bacterium. Thus the streptomycin resistance determined by some *E. coli* plasmids is due to the production of an enzyme which inactivates streptomycin, whereas the only kind of streptomycin resistance known to be due to a chromosomal gene in any bacterial species is that based on an alteration in a ribosomal protein component so as to make protein synthesis resistant to the drug.

The analogies between plasmid and virus genome are strong. To begin with, the replicative autonomy enjoyed by a plasmid is suggestive of a virus. There is also a strong analogy between the pili determined by transmissible plasmids and the protein coats of certain simple bacteriophages. The similarity in structure and dimensions between a pilus, which is a hollow tube built from a helically disposed array of identical protein subunits, and the protein coat of several of the filamentous phages is striking. Pilus-specific filamentous phages characteristically attach end-on to the pilus of the host and are then not easy to distinguish from the pilus itself except by specific affinity for antibodies or other phages (Fig. 36).

Even more suggestive is the mode of release from the host cell of filamentous phages; they are extruded lengthwise rather like a pilus from the cell surface, rather than being released by cell lysis. They contain single-stranded DNA and it is tempting to suppose that the injection of their nucleic acid from their tubular protein coat into the host cell shares a common mechanism with the cell-to-cell transfer of single-stranded plasmid or chromosomal DNA through the pilus. Following this line of thought it is possible to regard the pilus itself as the protein coat of a special kind of filamentous phage, adapted to the transfer of the plasmid nucleic acid from cell to cell while retaining contact throughout with the donor cell surface. The difference between a transmissible plasmid and a filamentous phage may be merely that the encapsulation of the infective nucleic acid into a protein coat and the injection of the nucleic acid from the protein coat into a fresh host cell, events which in the typical phage are separated in time and place, are, in the case of the transmissible plasmid, telescoped into one continuous process. The advantage from the phage's point of view of this change of habit is that a transmissible plasmid can acquire and transfer a longer stretch of DNA than can be held within the protein filament at a given instant whereas, in a typical phage, everything has to be accommodated within the infective particle. So long as it retains the DNA segment necessary for initiation of replication and the genes, including the pilus-protein gene, necessary for cell-to-cell transfer, a plasmid can 'pick up' chromosomal material with little reduction of its capacity for multiplication and infection. Indeed, a large increase in size brings a penalty in the form of increased time required for replication and transmission, but the extra genetic material brings compensating advantages to the plasmid to the extent that it improves the chances of survival and multiplication of the host cell.

Plasmids may change their properties not only by acquiring bacterial chromosomal genes but also by recombination with each other. Just as phage recombinants can arise from a mixed infection, so the co-existence in the same cell of two different compatible plasmids can result in the formation of new plasmid types combining genes from both. In some such cases the new types have been shown to have closed circular DNA of a size consistent with the conversion of two smaller circles into one larger one by a single cross-over event.

We may, then, picture plasmids as itinerant peddlers of off-the-peg genes, flourishing to the extent to which these genes satisfy the patrons on which they live, and now and again acquiring and offering for trial new lines of genetic merchandise, some of which may have originated from sources quite remote from the current bacterial host. Some of these exotic genes may ultimately become integrated into the chromosome and thus become part of the permanent genetic equipment of the bacterium.

The relation between bacterial cell and plasmid is indeed a subtle and complex one, with the plasmid having some of the aspects of a parasite, some of a symbiont and some of an integral component of the bacterial genetic system. The analogy with the relationship, on a higher level of organisation, between the eukaryotic cell and its plastids and mitochondria, organelles which are now widely believed to have been of prokaryotic origin, is a striking one.

8 Physico-chemical Mapping and Manipulation of the Genome

Perhaps the crowning achievement of classical genetics has been the detailed mapping of what are, after all, physical objects—the chromosomes and genes—without the use of any physical data. Everything was done by the classification and counting of progeny of controlled crosses, the classification often being made on the basis of relatively crude qualitative criteria. As we saw in Chapter 3, a great deal may be learned about the spatial relationships and internal structure of genes without the genes themselves ever being observed or handled as physico-chemical objects. Satisfying as this has been to geneticists, it may seem, to physical scientists, rather too good to be true.

During the last five years, however, a new set of methods has been developed, parallel with and complementary to classical genetic analysis, which enable the validity of the geneticists' deductions to be demonstrated to even the most sceptical chemist. With these new resources it is possible to map at least relatively simple genomes by physical methods and even to see and isolate single genes.

So far, the physico-chemical approach has been most successful applied to plasmids and viruses, especially bacteriophages, and there is heavy emphasis in this chapter on the *E. coli* phage *lambda*, which is now probably the best understood genetic system of all. In the future the same or analogous methods may prove applicable to the genomes of eukaryotic organisms, and what is called 'genetic engineering' may become a reality in higher plants and animals including man.

HYBRIDISATION OF NUCLEIC ACID MOLECULES

The infinitely varied nucleotide sequences shown by nucleic acid molecules, though of supreme biological importance, give little to the chemist to work on. Short of complete sequence determination—a practical impossibility with molecules tens of thousands of nucleotides in length—there is no chemical procedure for distinguishing one molecule from another with a different detailed sequence but similar overall composition. Fortunately, however, the complementary strands of double-stranded DNA, or the complementary DNA template and RNA transcript, can recognise *each other* with extreme accuracy, and this natural complementarity allows the biochemist to devise assays for specific sequences.

It was J. Marmur and his colleagues at Harvard University who first exploited DNA-DNA complementarity as a means for recognising homologies between DNA molecules. They showed that DNA which had been 'melted' to single strands by heating would spontaneously reform the original duplex structure on slow cooling under appropriate conditions of pH and salt concentration (a process called 'annealing'). They further showed that single-stranded DNA from different bacterial strains or species would form hybrid duplex molecules to an extent depending on the degree of relatedness of the organisms. Single-stranded molecules which are mis-matched to the extent of more than 5 or 10% of their bases form duplexes of appreciably reduced thermal stability, while if the degree of non-homology is much higher than this, hybrid duplexes are not formed at all. This principle has been widely used as an aid to assessing degrees of relatedness between organisms.

DNA-DNA hybridisation has acquired still greater importance since the development of electron microscope techniques for actually visualising hybrid DNA molecules, so that not only the overall degree of homology but the exact limits of homologous and non-homologous regions can be defined. This aspect is dealt with in the next section.

Of equal importance to DNA-DNA hybridisation is the hybridisation, first demonstrated by S. Spiegelman and B.D. Hall, of DNA with complementary RNA. A duplex molecule formed from one strand of DNA and a complementary strand of RNA has a fair degree of stability, though it is not as stable as a purely DNA duplex. If one can isolate a particular kind of RNA molecule (for example, ribosomal RNA, transfer RNA of specific kinds or specific messenger molecules) one can use it to identify the complementary DNA strand from which it was transcribed. There are various ways of annealing single-stranded DNA to complementary RNA, but perhaps the most convenient involves the absorption of the DNA to a nitrocellulose filter disc and then incubating the filter with radioactively labelled RNA in a salt solution buffered at neutral pH and held at a temperature above that necessary to keep the RNA in extended form (i.e. free of double-stranded structure due to self-annealing). Complementary RNA becomes bound to the DNA, unpaired single-stranded regions can be digested away by treatment with pancreatic ribonuclease, and the quantity of RNA held in stable double-stranded association with the DNA on the filter measured as radioactivity in a liquid scintillation counter. This technique is suitable for showing whether DNA-RNA hybrids are formed, and in what quantity, but the hybrids cannot be recovered from the filters. If it is necessary to study the hybrids under the electron microscope, it is necessary to carry out the annealing reaction in free solution, to separate the hybrids by some suitable method (i.e. by centrifugation in a density gradient) and to prepare the molecules for electron microscopy by the Kleinschmidt technique, already referred to in the last chapter (p. 112) and described below.

A more recent refinement uses the more stable annealing of complementary DNA strands to explore DNA-RNA homologies. The enzyme *reverse transcriptase,* recently found to be associated with many animal RNA viruses, synthesises DNA using a single-stranded RNA template—an inversion of the previously accepted rule for direction of transcription. Using this enzyme one can make a specific DNA transcript of any RNA molecule which it is possible to isolate in pure form. This transcript can then be used as a 'probe' for recognising complementary DNA sequences from which (or to which) the RNA could have been transcribed. This technique has recently been used for detecting DNA complementary to virus RNA in the nuclei of animal and human cells. The hypothesis, now apparently established in a number of cases, is that some RNA animal viruses can exist in latent form in cells through the transcription of their genetic information into DNA and the integration of the latter into the host chromosomes.

PHYSICAL MAPPING USING THE ELECTRON MICROSCOPE

In Chapter 2 mention was made of the use by Cairns of tritium labelling and autoradiography for tracing the outline of a large loop of DNA representing the entire *E. coli* genome. Cairns' studies required only the light microscope and, indeed, the extended *E. coli* chromosome is too large an object to be easily prepared for electron microscopy. Smaller DNA molecules, especially those of bacteriophages and plasmids, have, however, been studied with great success in recent years with the electron microscope.

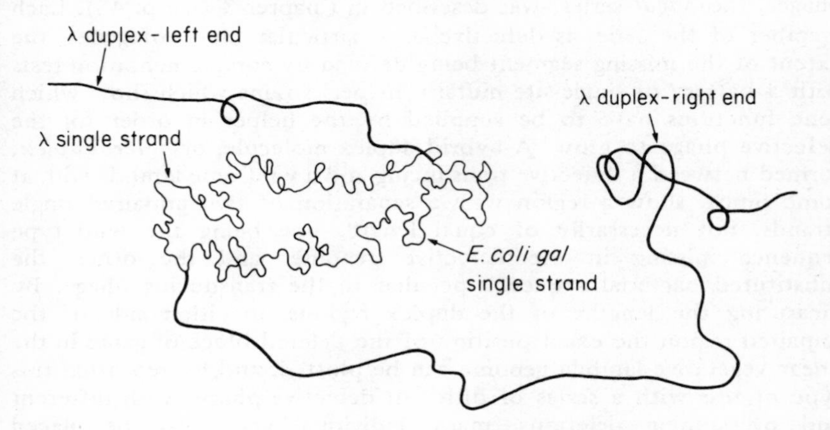

FIGURE 37 Electron micrograph of a λ/λd*gal* hybrid DNA molecule prepared by the Kleinschmidt method with 50% formamide. The limit of the deletion of λDNA in the λ*dgal* strand is obtained by measuring the length of the duplex from the end of the molecule to the unpaired single-stranded region. After Davis, R. W. and Parkinson, J. S. (1971) *J. Mol. Biol.* 56, 403.

Cairns, of course, was looking not at the DNA itself but at the silver grains deposited in the radiosensitive film as a result of the disintegration of the tritium atoms in the DNA. In order to be easily seen even under the electron microscope DNA molecules have to be made thicker and more opaque to electrons. The necessary thickening is accomplished through a technique devised by Kleinschmidt in which the specimen is mixed with cytochrome c, a readily available and markedly basic protein which binds in a thick coat to the acidic DNA. When spread on electron microscope grids and 'shadowed' or stained with electron-dense heavy atoms such as platinum, palladium or uranium, the nucleic acid-protein fibres can be photographed with diagrammatic clarity.

As an additional bonus it is possible to distinguish between single- and double-stranded DNA. In aqueous media single-stranded regions collapse on themselves to form compact aggregates or 'bushes', but in a medium containing 50% formamide single strands become more or less extended and their lengths can be measured; they can be clearly distinguished from double-stranded regions by their thinner appearance and their kinky rather than rod-like contours (see Fig. 37).

The distinction between single- and double-stranded regions has been the key to the combined physical and genetic mapping of several bacteriophage genomes. A very powerful method involves the formation of hybrid molecules by the annealing of complementary single strands from two phages, one with a deletion of part of the genome containing a defined block of genes. In *lambda* bacteriophage particularly, there is an abundance of defective strains isolated as specialised transducing particles. The origin of one class of defective transducing phages, the λ*dgal* series, was described in Chapter 3 (see p. 43). Each member of the series is defective in a particular block of genes, the extent of the missing segment being defined by complementation tests with a battery of single-site mutant 'helper' strains which show which gene functions have to be supplied by the helper in order for the defective phage to grow. A hybrid duplex molecule, or *heteroduplex*, formed between a defective transducing and a wild type lambda will, at some point, show a region with a separation of two unpaired single strands, not necessarily of equal length, one being the wild type sequence missing in the defective genome, and the other, the substituted bacterial sequence peculiar to the transducing phage. By measuring the lengths of the duplex regions on either side of the unpaired region the exact position of the deleted block of genes in the linear vegetative lambda genome can be plotted, and by repeating this type of test with a series of different defective phages with different and overlapping deletions, many individual genes can be placed accurately in the physical DNA molecule as well as in the genetically deduced map. Fig. 38 shows some of the results of this kind of analysis.

One difficulty with the method just outlined is that it is not apparent under the electron microscope which end of a DNA molecule is which. Thus it may, for example, be evident that a given deletion is

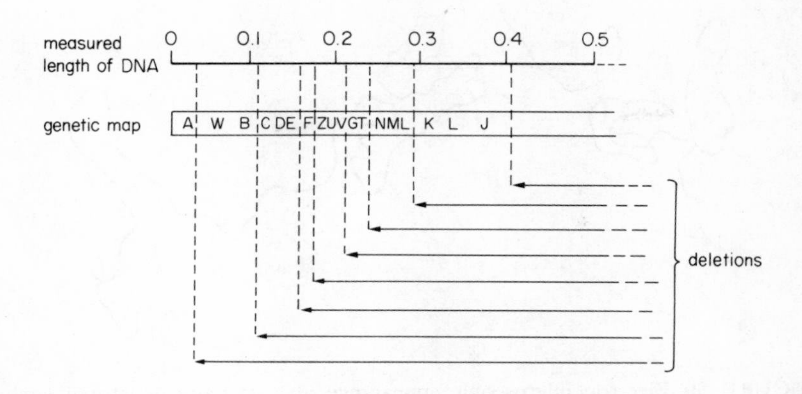

FIGURE 38 Physical map of the left end of the λ genome. The diagram shows, from top to bottom, (a) the distance along the DNA molecule, as a fraction of the total length, (b) the positions of genes, symbolised by letters and (c) deletions in various λd*gal* phages, showing both the block of genes missing in each case (determined by complementation tests with helper phages) and the measured extent of the deletion as seen with the electron microscope in a preparation such as that shown in Fig. 37.

located 20% of the way along the genome from one end, but the question of *which* end may sometimes remain unanswered. This problem can be overcome in principle by introducing physical 'landmarks' which can serve to distinguish one end of the DNA from the other. For example, instead of annealing the unknown deletion DNA with wild type DNA, it may be hybridised with a DNA carrying a deletion at a known position. The lambda deletion mutant b_2 has been frequently used as a reference marker for physical mapping (see Fig. 40b). Landmarks may also be introduced by the technique of *denaturation mapping*, which has been applied successfully to *lambda* and T7 bacteriophages as well as to the well-studied animal virus SV40 (simian virus 40). The idea is to heat the DNA preparation, or to subject it to alkali treatment, to an extent sufficient to break down only the less stable hydrogen-bonded stretches of duplex structure. The G-C base pairs are somewhat more stable than the A-T pairs, supposedly because they are held together by three hydrogen bonds rather than two. Thus a partially denatured DNA duplex tends to show single-stranded structure in those regions richer in A-T and poorer in G-C. A particular type of DNA molecule displays a reproducible and characteristic pattern of single-stranded loops if spread for electron microscopy in formamide. Fig. 39 illustrates the typical pattern shown by lambda DNA.

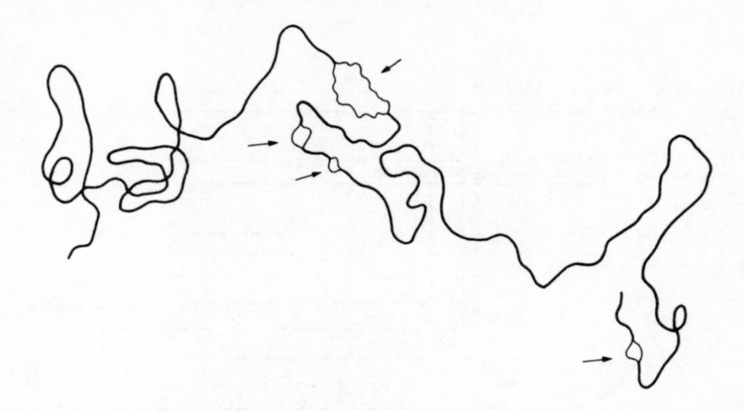

FIGURE 39 Electron microscopic appearance of a partially denatured lambda DNA molecule, prepared by the Kleinschmidt method with 50% formamide. The exact positions of the single-stranded regions (arrowed) vary from one molecule to another, but occur with highest probability in the regions indicated, respectively about 50, 75 and 99.5% of a molecule-length from the left hand end of the molecule (equivalent to the left-hand end of the map – cf. Fig. 40.) After Inman, R.B. and Schnös, M. (1970) *J. Mol. Biol.*, 49, 93.

SPECIFIC ENZYMIC FRAGMENTATION OF DNA AND ORDERING OF THE FRAGMENTS

As we have just seen, physical mapping of considerable refinement can be carried out on the genome of a medium-sized virus, such as lambda bacteriophage. Lambda DNA, however, is still rather large (about 46 500 base pairs) and for many purposes it is desirable to have something smaller to work with. It is, in fact, possible to cut up the DNA of lambda, and that of many other viruses, into pieces of more manageable size with specific endonucleases—enzymes that will cleave DNA molecules at internal positions rather than degrading them from the ends. Biochemists now have at their disposal an array of endonucleases produced by bacteria and bacterial plasmids. The function of these enzymes seems to be one of 'restriction', that is the prevention of the establishment of foreign DNA in the cell. The resident genome protects itself against its own restriction enzymes by modifying, usually by methylation, those of its own base sequences which would otherwise be vulnerable. The nature of the specificity of many restriction endonucleases is known, and in several cases it involves the simultaneous cleavage of both DNA strands in somewhat staggered positions which are symmetrically related. For example, the endonuclease determined by the *E. coli* resistance-determining plasmid RI attacks the double-stranded sequence shown:

$$\begin{array}{c} \downarrow \\ 5'\text{G-A-A-T-T-C}3' \\ 3'\text{C-T-T-A-A-G}5' \\ \uparrow \end{array}$$

nucleotide linkages being cleaved in the arrowed positions. Such a specific symmetrical sequence is likely to occur only a few times in 10^5 base pairs, and restriction endonucleases can thus often be used to reduce a large DNA molecule to a limited number of reproducible fragments which can then in principle be purified and studied separately.

One of the best examples of specific nuclease cleavage is the use of the RI endonuclease, referred to above, to reduce the lambda genome to six separable pieces. These pieces are all different sizes and they can be separated from one another by electrophoresis through a gel made of polymerised acrylamide. The polyacrylamide exercises a stringent sieving action, slowing the movement of the fragments in the electric field to an extent dependent on their respective sizes. The positions in

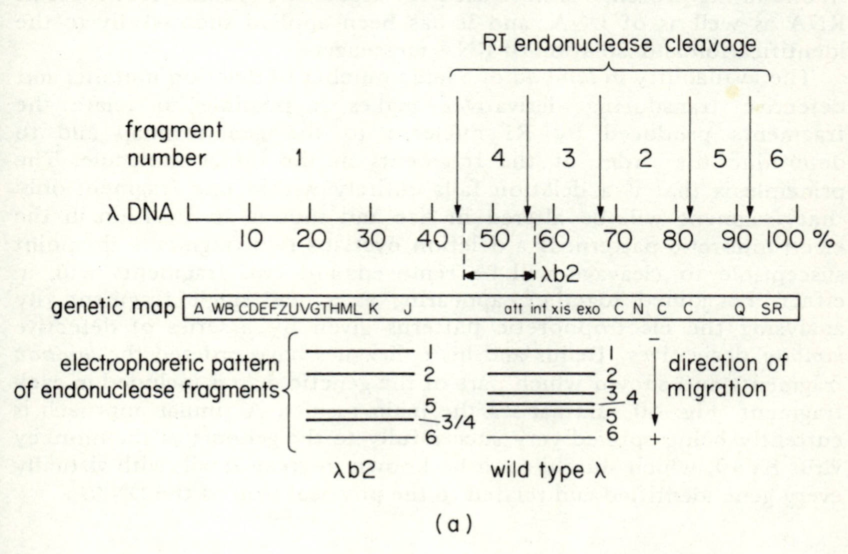

FIGURE 40(a) The splitting of the λ genome into fragments by RI endonuclease, the separation of the fragments by size by electrophoresis in polyacrylamide gel, and the principle of the use of deletion mutants to establish the sequence of the fragments. The deletion mutant λb_2, which lacks the attachment region *att*, and hence cannot integrate into the host chromosome, is shown here as one example. The disappearance in λb_2 of fragments 2 and 3, and the appearance of a new fragment (presumably comprising the left end of 4 joined to the right end of 3), is taken to show that 2 and 3 are adjacent with the cleavage site between them deleted in λb_2. The sequence of all six fragments was established in a similar way using a series of other deletions. Abridged from Allet, B., Jeppensen, P.G.N., Katagiri, K.J. and Delius, H. (1973), *Nature*, 241, 120.

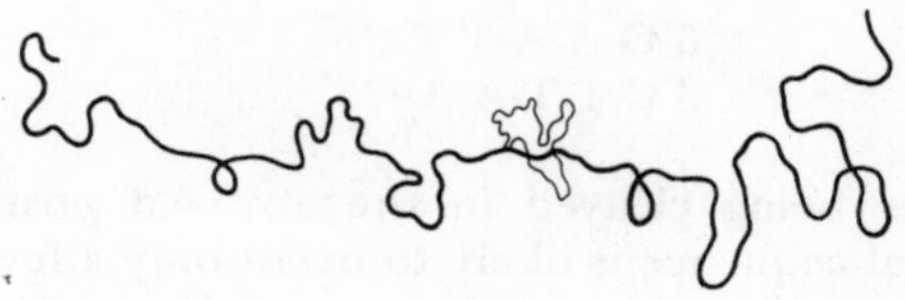

(b)

FIGURE 40(b) The appearance of the wild type/b_2 heteroduplex.

the gel of the different fragments can be shown by staining them with methylene blue, and the separated fragments can then be cut out of the gel and recovered in pure form. In passing we may note that polyacrylamide gel electrophoresis, a technique first developed for fractionating proteins, is now used for separating specific sequences of RNA as well as of DNA, and it has been applied successfully to the identification and isolation of RNA messengers.

The availability in *lambda* of a large number of deletion mutants and defective transducing derivatives makes it possible to relate the fragments produced by RI nuclease to the genetic map and to determine the order of the fragments in the intact molecule. The principle is that if a deletion falls entirely within one fragment only that fragment will be altered in size and thus in its position in the electrophoretic pattern. If a deletion overlaps two fragments the point susceptible to cleavage will be removed and two fragments will, in effect, be joined together, appearing as a new single fragment. By analysing the electrophoretic patterns given by a series of defective *lambda* derivatives, Delius and his colleagues have ordered the *lambda* fragments and shown which part of the genetic map is included in each fragment. Fig. 40 summarises the main results. A similar approach is currently being applied very successfully to the genome of the monkey virus SV40, which should soon be known in great detail, with virtually every gene identified and related to the physical map of the DNA.

POSSIBILITIES OF PHYSICAL FRACTIONATION OF THE EUKARYOTE GENOME

Any detailed physical and chemical analysis of DNA depends on the isolation of relatively short sequences in pure form. As we have already seen, much can be done with the small genomes of plasmids and some of the smaller viruses, but even so fractionation into still smaller pieces is desirable for many purposes. When one comes to consider the much more complex *E. coli* genome or the very much larger amounts of DNA in eukaryote chromosomes, it is clear that it is only by fractionation that any progress can be made at all.

Up to the present, success in fractionating the DNA of higher eukaryotes has been very limited. Different fractions with markedly different proportions of A-T to G-C base pairs differ in buoyant density (the more G-C the higher the density) and can be separated on this basis in caesium chloride density gradients (cf. p. 16). On the whole, different parts of a complex genome tend to resemble one another in overall base composition, but exceptions to this rule are the 'satellite' DNA fractions of many animals, so-called because they form minor bands in the ultracentrifuge separable on the basis of buoyant density from the main band DNA. In several mammals this satellite consists of extremely simple and highly repetitious sequences which do not look as if they can contain any significant genetic information and may serve some kind of structural function in the chromosome. More interesting is the satellite of high G-C content found in the African horned toad, *Xenopus laevis*, which turns out to be many hundreds of repeated copies of the genes which are transcribed into 28s and 18s ribosomal RNA. Only very highly repetitious genes, of which the rRNA genes are the prime example, are likely to be isolable from bulk DNA by physical methods. Most genes are probably present in the genome as single copies and the only way of isolating them that one can see in principle (though not yet in practice) is through trapping the single-stranded DNA fragments containing them by hybridization with specific complementary messenger RNA. Pure messengers can now be obtained in favourable circumstances (for example, haemoglobin messenger can be isolated from the highly specialised haemoglobin-producing blood cells), and this approach to gene isolation is certainly feasible in some cases, though more usually the messenger is unlikely to be obtainable in sufficient quantity.

The genomes of eukaryotes are, of course, naturally segmented into chromosomes which, since they are of different sizes, could in principle be fractionated on the basis of sedimentation velocity in the centrifuge. Some limited success has been obtained in separating the highly contracted metaphase chromosomes of animal cells into size classes, though not into single chromosome classes. The best hope for isolating pure chromosomes is probably in yeast (*Saccharomyces cerevisiae*), where Petes has shown that the nuclear DNA can be isolated in the form of unbroken linear DNA molecules of different sizes, each one corresponding to one chromosome. These molecules can be visualised and measured using the electron microscope, and they are each considerably smaller than the DNA loop of *E. coli*. At the time of writing methods have not been worked out for isolating any one yeast chromosomal DNA molecule from the 15 or so others. Such a molecule, though considerably larger than the virus genomes which are now being worked upon (see below), could be small enough to present a realistic challenge to the nucleic acid chemist.

VISUALISATION OF SPECIFIC GENES

Use of gene transcripts as labels

Having isolated a specific small fragment of DNA it is possible in principle to determine whether the fragment includes a particular gene by determining whether it will anneal with the RNA transcript of the gene. Provided one can obtain the RNA transcript—rRNA, tRNA or messenger RNA as the case may be—it is relatively easy to test for the presence of a complementary DNA strand by labelling the RNA with a radioactive isotope and looking for the appearance of radioactivity in DNA/RNA hybrid. More than this, it is sometimes possible actually to see the RNA annealed to the DNA strand with the electron microscope

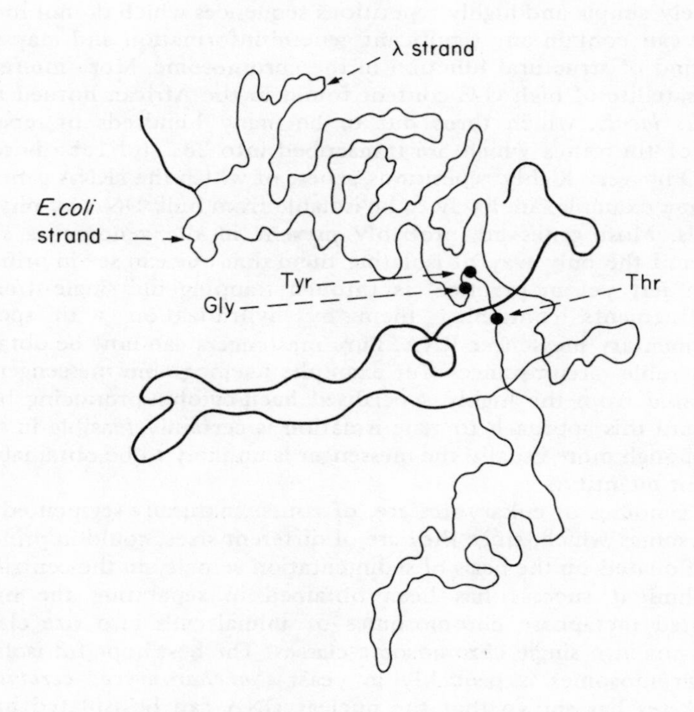

FIGURE 41 Visualisation of three *E.coli* transfer RNA genes (for tyrosine, glycine and threonine respectively). The electron micrograph illustrated shows a heteroduplex formed between an essentially wild type lambda strand and a specialised transducing lambda strand in which over 40% of the sequence is replaced by a sequence of *E.coli* DNA including the three genes. The dense blobs are annealed ferritin-labelled tRNA molecules of the three kinds; it is known which is which by accurate measurements of the positions of the three tRNA's annealed separately. From Wu, M., Davidson, N. and Carbon, J. (1973) *J. Mol. Biol.* 78, 23-24.

and to determine exactly where in the DNA fragment under study the gene in question is located.

This approach to direct visualisation of genes has, at the time of writing, been applied only in a few cases but two notable examples may be mentioned. The first involves some of the most readily accessible gene transcripts, the relatively abundant and easily isolated transfer RNA molecules. These RNA's, being only 80 or so nucleotides long, would be difficult to pick out in an electron micrograph unless labelled with some conspicuously electron-dense marker. A convenient marker is ferritin, a small protein containing an atom of iron. Wu, Davidson and Carbon succeeded in coupling ferritin to various tRNA molecules without impairing the ability of the the latter to form a duplex with complementary DNA. The study made use of a number of lambda and ϕ80 derivatives carrying suppressor bacterial genes coding for various modified tRNA molecules (cf. p. 90). Separated single DNA strands from these phages, when annealed with complementary strands from normal phage, showed single-stranded loops corresponding to the bacterial DNA which they carried (cf. Fig. 37). When the single-stranded loop was the strand from which the tRNA was transcribed it was possible to anneal the tRNA to it and see its exact position as an electron-dense blob due to the ferritin molecule. This procedure was used to order the genes for three different tRNA species (for glycine, threonine and tyrosine respectively); these genes turned out to be very closely linked—separated in fact, by only 1200 nucleotides (Fig. 41). As well as demonstrating the relative positions of the three genes, the experiments also established which of the two DNA strands of each transducing phage was the one used for transcription.

The other example concerns the bacteriophage T7. This phage has been the object of increasing attention during the last few years, one main attraction being its small size and relative simplicity. It is thought to have only 17 genes, and the messenger RNA transcripts, as well as the polypeptide products, corresponding to practically all of them have been identified by electrophoretic fractionation of extracts of infected bacteria of polyacrylamide gels (essentially the same method as used for DNA fragments—see p. 127). Mutations in gene number 1 affect the T7-encoded RNA polymerase, which is produced early in the course of infection and is a prerequisite for the transcription of most of the other phage genes. The messenger RNA corresponding to gene 1 was recovered from the electrophoretic gels by Hyman and Summers, and this isolated and purified RNA was shown to anneal specifically with one of the two separated single strands of T7 DNA—the arbitrarily designated 'right' strand. It was possible, using an appropriate modification of the Kleinschmidt technique, actually to see the DNA/RNA duplex covering a section of the otherwise single-stranded DNA and to determine that it occupied the space betwen points 8 and 17% of the distance along the linear genome. The annealed RNA was seen rather more clearly in further experiments in which the complementary 'right' and 'left' single strands of the T7 DNA were reannealed in the presence of an excess of the gene 1 messenger; the RNA interposed itself between the DNA strands, displacing a single-stranded DNA loop which marked the position of the gene. As other messengers are identified and isolated other genes will no doubt be visually mapped in the same way.

Finally, the use of messenger RNA to make the locus of a gene visible under the light microscope must be mentioned. Of all eukaryote messenger RNA's, that for the globin component of haemoglobin is the most easily available, for it is produced by reticulocytes (immature red blood cells) almost to the exclusion of other messengers and can be recovered in virtually pure form from reticulocyte polysomes. Hirschorn and his colleagues obtained this messenger RNA labelled with tritium and annealed it to the partially denatured DNA of human metaphase chromosome spread on a microscope slide. The position of the radioactivity was shown by autoradiography (cf. p. 18) and, as expected, it was found to be restricted to two loci on different chromosome pairs, presumably corresponding to the genes for the globin α and β chains.

GENE ISOLATION

The isolation in pure form of single genes or small groups of genes is now entirely feasible, and has actually been accomplished in a few instances. The examples available at the time of writing are all in *Escherichia coli* and involve the use of specific transducing phages (derivatives either of lambda or of $\phi 80$) which, in as much as they each carry a very small and specific fragment of the bacterial genome, have already accomplished most of the necessary fractionation. Starting with such phages, all that is necessary is to devise a method for separating the desired segment of bacterial DNA from the rest of the DNA of the transducing phage.

Such a method was devised by Shapiro and others at Harvard University, and was first applied by them to the isolation of the genes of the *lac* operon. The principle was as follows. The two complementary strands of the DNA of lambda or of $\phi 80$ differ from one another sufficiently in overall base composition to be separated as heavy (H) and light (L) bands in a caesium chloride density gradient in the ultracentrifuge (cf. p. 16). The separation is improved by the inclusion in the solution of the synthetic RNA polymer poly-UG, which binds more to one strand than to the other. Duplexes can only be reformed by mixing H and L; pure H or pure L strands are not self-complementary and remain single-stranded. The experiment was made possible by the availability of two transducing phages in which the *lac* operon was carried in opposite orientations. Calling the two complementary strands of the operon DNA 'left' and 'right', in one phage the 'left' strand was integrated in the H strand of the phage and the 'right' strand into the L strand, while in the other phage 'left' was in L and 'right' in H. The DNA of each transducing phage was separated into its H and L strands, and the two kinds of H strand, one from each phage, were annealed. The only complementary sequences in this mixture were the 'right' and 'left' strands of the *lac* operon, and so the annealing resulted in the formation of double-stranded *lac* DNA flanked on each side by long single-stranded 'tails' of similar rather than complementary

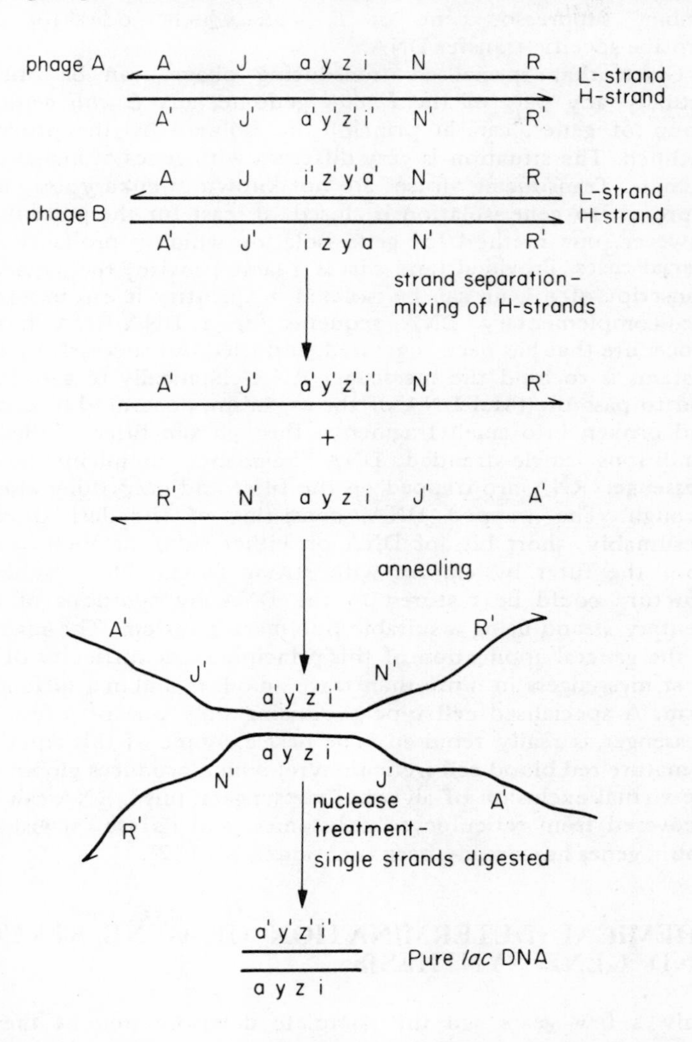

FIGURE 42 The principle of the method used for the purification of the *lac* operon of *E. coli*, from two transducing phages carrying the *lac* segment in opposite orientations. *A, J, R, N,* are *lambda* genes and *a, y, z, i* are genes of the *lac* operon (cf. Figs. 30 and 40). The complementary sequences of the double-stranded DNA of each gene are represented by the same letter primed (e.g. A′) and non-primed (e.g. A).

sequence. These single-stranded 'tails' could be digested away by a nuclease enzyme to which double-stranded DNA was resistant, leaving pure *lac* operon as the only macromolecule in the preparation. Fig. 42 illustrates the principle of the method. Since its first description essentially the same method has been used for the isolation ofthe 'amber' suppressor gene of *E. coli* which codes for a mutant tyrosine-specific transfer DNA.

Given that specialised transducing phages can be obtained for virtually any part of the *E. coli* genome, any *E. coli* gene, or short group of genes, can in principle be isolated by the procedure just outlined. The situation is very different with genes of higher plants and animals. Transducing viruses are not known in eukaryotes, and so this approach to gene isolation is closed, at least for the present. There is, however, one method for gene isolation which is probably feasible in special cases. Provided (and this is a large proviso) the messenger RNA transcript of a gene can be isolated in quantity it can be used to trap the complementary DNA sequence as a DNA/RNA hybrid. One procedure that has been suggested, and tried out successfully in a model system, is to bind the messenger RNA chemically to a cellulose filter and to pass the total DNA of the organism, denatured to single strands and broken into small fragments, through the filter. Under the right conditions single-stranded DNA fragments complementary to the messenger RNA are trapped on the filter and everything else will pass through. The trapped DNA, consisting of the desired gene with, presumably, short bits of DNA on either side, can then be recovered from the filter by elution with strong formamide. Double-stranded structure could be restored to the DNA by synthesis of a complementary strand using a suitable polymerase system. The main obstacle to the general application of this principle is the difficulty of obtaining most messengers in more than trace amounts and in a sufficiently pure form. A specialised cell type producing only one or a few species of messenger is really required. The best example of this situation is the immature red blood cell (reticulocyte) which produces globin mRNA to the virtual exclusion of all other messengers; this mRNA can be readily recovered from reticulocyte polysomes, and its use for visualising the globin genes has already been mentioned (p. 132).

CHEMICAL DETERMINATION OF GENE STRUCTURE, AND GENE SYNTHESIS

Only a few years ago the complete determination of the chemical structure of a gene seemed an almost unattainable goal. Today, however, not only has very impressive progress been made in the determination of the base-pair sequences of substantial sections of a few genes, but one gene has even been synthesised.

Since the techniques available for sequence determination of RNA are at present easier and more effective for those for DNA, the readiest approach to the structure of the gene is often through the sequence of the RNA transcribed from it. An RNA transcript can in principle be obtained either as a molecule (ribosomal RNA, transfer RNA or messenger RNA) made naturally in a living cell and purified from the cell extract, or as the product of the action of RNA polymerase using the gene sequence as a template in a cell-free system. Each of these approaches has been used successfully. The complete nucleotide sequences of many different tRNA's are now known, as is that of 5S ribosomal RNA of *E. coli*; substantial progress has also been made in the sequence determination of the larger 18S and 28S ribosomal RNA species. Once these RNA sequences have been established we automatically know what must be the structures of the DNA segments from which they are transcribed, but it remains to show where in the genome these deduced genes are. As was mentioned above, if one has an approximate idea of where to look it is sometimes possible to observe, under the electron microscope, the annealing of the RNA transcript with the DNA sequence from which it was presumably transcribed. In such a case one knows both the gene structure and, within close limits, its location in the longer DNA sequence of the chromosome.

The alternative method—the sequencing of artificial transcripts—has recently been used to great effect in the determination of the sequence of one region of the *E. coli lac* operon (cf. pp. 99-101), namely the operator region which binds specifically to the repressor protein—the product of the *lacI* gene. The results of this work were outlined in Chapter 6 (p. 101, and Fig. 31). Such use of artificial cell-free transcripts of DNA segments overlapping the boundaries of genes can give vital information not necessarily obtainable from natural gene transcripts, still less from the amino-acid sequences of the protein translation products of RNA messengers, concerning the sequences in DNA which provide signals for the initiation and termination of transcription.

One of the classical tenets of organic chemistry is that analysis should be confirmed by synthesis. The wholly synthetic gene, which for so long remained a remote fantasy, was realised in 1972 by a team of 14 headed by H.G. Khorana. The gene chosen for synthesis was one of the smallest known—that for alanine-specific transfer RNA of yeast. The sequence of 77 base pairs in this gene was known from the corresponding single-strand sequence in the tRNA, and it was put together through a long series of operations of great ingenuity, the details of which are beyond the scope of this book. Put in the most simple way, the strategy adopted was to synthesise the chain piecemeal in the form of overlapping complementary single-stranded sequences, to anneal them to form double-stranded chains with single-stranded gaps, and then to complete the double-stranded structure by the joining activity of DNA ligase.

Another approach, which may be successful as a means of synthesis of at least part of a gene, is to use the single-stranded RNA transcript of the gene as a template for the activity of reverse transcriptase (already mentioned on p. 123). Two groups of workers, headed respectively by

Baltimore and by Spiegelman, have tried out this method using globin messenger RNA, isolated from reticulocyte polyribosomes, for the synthesis of a single-stranded DNA which was shown to be complementary to and about the same size as the messenger. This DNA will be nearly equivalent to one strand of the globin gene (or rather of each of the two globin genes, one for the β and one for the α chain of haemoglobin), but it may not be quite complete since the reverse transcriptase may well start its copying some way into the sequence, missing out some nucleotides near one end.

The ultimate *tour de force*, the introduction of the synthetic gene into a living cell and the demonstration that it is there able to replicate and be transcribed into the normal RNA product, has not yet been reported, though the technical means appear to be available, as we shall see from the following and final section.

It should be remembered that, in order to function in a cell, a gene must possess the sequence of nucleotides which furnishes the signal for the commencement of transcription. If this sequence is not itself transcribed in its entirety, as it may quite possibly not be, the transcript will lack an essential part of the information necessary for the functioning of the gene, even though it does contain all the information necessary for the functioning of the gene product. So in order to make genes that will actually function when introduced into a cell, more needs to be known about the nature of the transcription-initiation signal, and whether it is itself transcribed or not.

THE CONSTRUCTION OF ARTIFICIAL GENETIC VECTORS

We have already dwelt on transducing bacteriophages in connection with use in the isolation of specific segments of bacterial DNA. Perhaps equally momentous is the realisation that they provide a general model for a procedure for the introduction and integration of artificial or foreign genes into cells. In naturally produced transducing phage the bacterial DNA gets into the phage genome as the result of an aberrant excision from the integrated state, due to an 'illegitimate' crossover (Fig. 18). As a result of the new methods for manipulating DNA molecules with specific enzymes, it is now possible to insert any desired small piece of DNA into a virus or plasmid genome artificially. It may then be possible to use the augmented or partially substituted virus genome as an artificial vector, carrying the foreign DNA segment into the chromosome of the recipient cell. This type of operation is a very immediate possibility using bacteria and bacteriophages; the applicability of the method to cells of higher organisms depends on whether suitable animal viruses, capable of integrating into the chromosomes, are available. I shall first describe in outline how compound DNA molecules may be used and then conclude with a brief discussion of the relevant animal viruses.

There are various procedures for the artificial joining of DNA molecules; the simplest involve fragments of DNA which have been produced by the action of endonucleases which, like RI endonuclease (cf. p. 127), cut the two strands at staggered and symmetrically related positions. Such fragments are naturally 'sticky' because of the tendency to anneal of the complementary single-stranded sequences at their ends. Thus an RI endonuclease fragment would be able to form a circle, or two or more fragments could unite head-to-tail. Following annealing by hydrogen bonding, the joints can be covalently sealed by the enzyme DNA-ligase. Cohen and others have demonstrated the joining of RI fragments from two different resistance-transfer plasmids to form a closed-circular compound plasmid which proved to be capable of propagating itself after being introduced into *E. coli* cells.

A still more striking application of the technique, which has become known as *genome fusion*, was the recent creation of a derivative of an *E. coli* plasmid carrying a block of rRNA genes from the satellite DNA (see p. 129) of the African found toad *Xenopus laevis*. On being infected into *E. coli* this bizarre plasmid was able to replicate and was shown to cause the transcription of toad ribosomal RNA in the bacterial cell. This experiment opens up most exciting possibilities, for, by fusing lambda or plasmid RI endonuclease fragments with DNA fragments similarly generated from another organism (say yeast, *Neurospora* or the fruit-fly *Drosophila*) it should be possible to fuse virtually any eukaryotic gene into a DNA molecule which can replicate in *E. coli*. Enormous numbers of different bits of DNA would, of course, be picked up by such a procedure, and the identification of any particular product would depend on the development of efficient methods for selecting bacteria with specific new gene activities. Whether eukaryotic genes can be translated as well as transcribed in prokaryotes, is not yet clear; it is possible that bacterial ribosomes will not recognize eukaryotic translation start signals. However, *if* this kind of experiment works, it will provide a means of selective propagation of eukaryotic genes, andthe study of their fine structure and modes of regulation, in a way that would otherwise hardly be possible.

There has also been speculation that techniques of genome fusion involving mammalian viruses could be used to repair genetically determined metabolic defects in man. If it were possible to fuse a human gene into a virus genome capable of integration into the human chromosome it might be possible to cure a genetic defect by transduction. Thus a congenital defect in a liver enzyme might be partially repaired by the implantation into the patient of some of his own liver cells which had been for a time cultured *in vitro* and treated with a suitable transducing virus. There are, in fact, several animal viruses, such as SV40 and polyoma, whose genomes are known to be capable of integration into chromosomes and which might be used in the synthesis of artificial gene vectors. This sort of 'genetic engineering', however, seems as yet a rather remote possibility and is fraught with hazards as well as with technical difficulties. Even leaving aside the question of the feasibility of introducing sufficient cured cells to make the treatment effective, there are still the difficult problems of isolating the relevant gene (complete with its own transcription signals) and, above all, of finding a safe virus to use as vector. This last consideration is particularly serious, since the animal viruses known to integrate into the hose DNA are often oncogenic (tumour inducing).

The possibility of *cloning* selected pieces of DNA from virtually any source by integrating them artificially into the genomes of viruses or plasmids has generated a high degree of scientific excitement, but has also aroused misgivings. The synthesis of new infective agents with not entirely predictable properties must involve some risk, particularly when the agents include segments of animal viruses, or consist of pieces of animal genomes (which may harbour latent virus DNA) fused into plasmids capable of spreading in a universally distributed enteric bacterium such as *Escherichia coli.* The question of establishing procedures which will allow the scientifically valuable work to proceed while avoiding serious risk is currently the subject of lively discussion within the scientific community (see the statement by P. Berg *et al.* and the report of the Ashby Committee, listed in the Bibliography pp. 142-3).

References and Further Reading

Chapter 2

Cairns, J. (1963). 'The bacterial chromosome and its mode of replication as seen by autoradiography', *J. Mol. Biol.*, 6, 208. (Direct visual evidence for the semi-conservative replication of DNA in *E. coli.*)

Fox, M.S. and Allen, M.K. (1964). 'On the mechanism of deoxyribonucleic acid integration in pneumococcal transformation'. *Proc. Nat. Acad. Sci., U.S.A.,* 52, 412. (Evidence for integration of single stranded fragments of DNA into recipient duplex molecules.)

Fraenkel-Conrat, H., Singer, B.A., and Williams, R.C. (1958). 'The nature of the progeny of virus reconstituted from protein and nucleic acid from different strains of tobacco mosaic virus', in *The Chemical Basis of Heredity* (edited by W.D. McElroy and B. Glass), Johns Hopkins Press.

Goulian, M., Kornberg, A. and Sinsheimer, R.L. (1967). 'Enzymatic synthesis of DNA, XXIV. Synthesis of infectious phage φX174 DNA. *Proc. Nat. Acad. Sci., U.S.A.* 58, 2321. (Demonstration of infectivity of phage DNA synthesized in a cell-free system).

Hershey, A.D., and Chase, M. (1952). 'Independent functions of viral protein and nucleic acid in growth of bacteriophage', *J. Gen. Physiol.,* 36, 39.

Meselson, M., and Stahl, F.W. (1958). 'The replication of DNA in *Escherichia coli*. *Proc. Nat. Acad. Sci., U.S.A.* 44, 671.

Robertus, J.D. *et al.* (1974). Structure of yeast phenylalanine tRNA at 3 A resolution. *Nature,* 250, 546. (The first complete 3-dimensional structure of an RNA molecule).

Schaeffer, P. (1964). 'Transformation'. In *The Bacteria*, vol. V (edited by I.C. Gunsalus and R.Y. Stainer), Academic Press. (A good general review.)

Watson, J.D., and Crick, F.H.C. (1953). 'A structure for deoxyribose nucleic acid', *Nature,* 171, 737.

Watson, J.D., and Crick, F.H.C. (1953). 'Genetical implications of the structure of deoxyribose nucleic acid', *Nature,* 171, 964.

Chapter 3

Benzer, S. (1957). 'The elementary units of heredity', in *The Chemical Basis of Heredity* (edited by W.D. McElroy and B. Glass), Johns Hopkins Press, Baltimore.

Dubnau, D., Goldthwaite, C., Smith I. and Marmur, J. (1967). 'Genetic mapping in *B. subtilis*'. *J. Mol. Biol.*, 27, 163. (Mapping both by transduction and by density transfer.)

Fincham, J.R.S. and Day, P.R. (1971). *Fungal Genetics.* Third edition, pp. 402. Blackwell Scientific publications, Oxford. (For genetic mapping in fungi.)

Jacob, F., and Wollman, E.L. (1961). *Sexuality and the Genetics of Bacteria,* pp. 374. Academic Press, New York and London.

Levinthal, C., and Visconti, N. (1953). 'Growth and recombination of bacterial viruses', *Genetics,* 38, 500.

Nester, E.W., Ganesan, A.T., and Lederberg, J. (1963). 'Effects of mechanical shear on genetic activity of *Bacillus subtilis* DNA', *Proc. Nat. Acad. Sci., Wash.,* 49, 61.

Ozeki, H. (1959). 'Chromosome fragments participating in transduction in *Salmonella typhimurium*', *Genetics,* 44, 457.

Petes, T.D.—see reference under Chapter 8.

Silver, S.D. (1963). 'The transfer of material during mating in *Escherichia còli*', *J. Mol. Biol.,* 6, 349.

Yanofsky, C., and Lennox, E.S. (1959). 'Transduction and recombination study of linkage relationships among the genes controlling tryptophan synthesis in *Escherichia coli*', *Virology,* 8, 425.

Zinder N.D. (1953). 'Infective heredity in bacteria', *Cold Spring Harbor Symp. Quant. Biol.,* 18, 261.

Chapter 4

Benzer, S. (1961). 'On the topography of genetic fine structure', *Proc. Nat. Acad. Sci., U.S.A.,* 47, 403.

Champe, S.P., and Benzer, S. (1962). 'Reversal of mutant phenotypes by 5-fluorouracil: an approach to nucleotide sequences in messenger-RNA', *Proc. Nat. Acad. Sci., U.S.A.,* 48, 532.

Davis, B.D. (1950). 'Studies on nutritionally deficient bacterial mutants isolated by means of penicillin', *Experientia,* 6, 41.

Drake, J.W. (1963). 'Properties of ultraviolet-induced *rII* mutants of bacteriophage T4', *J. Mol. Biol.,* 6, 268.

Freese, E. (1961a). 'The molecular mechanisms of mutations', *Proc. 5th Int. Congr. Biochem., Moscow.* Pergamon Press.

Freese, E. (1961b). 'The chemical and mutagenic specificity of hydroxylamine', *Proc. Nat. Acad. Sci., U.S.A.,* 47, 845.

Luria, S.E., and Delbrück, M. (1943). 'Mutations of bacteria from virus sensitivity to virus resistance', *Genetics,* 28, 49.

Lederberg, J., and Lederberg, E.M. (1952). 'Replica plating and indirect selection of bacterial mutants', *J. Bacteriol.*, 63, 399.

Strelzoff, E. (1961). 'Identification of base pairs involved in mutations induced by base analogues', *Biochem. Biophys. Res. Comm.*, 5, 384.

Yourno, J., Ino, J. and Kohno, T. (1971). 'A hotspot for spontaneous frameshift mutations in the histidinol dehydrogenase gene of *Salmonella typhinurium*'. *J. Mol. Biol.*, 62, 233. (see also *Nature*, 236, 338 (1972).

Chapter 5

Ames, B.N., and Hartman, P.E. (1963). 'The histidine operon', *Cold Spring Harbor Symp. Quant. Biol.*, 28, 349.

Benzer, S.—see Chapter 3 references.

Crick, F.H.C., Barnett, L., Brenner, S., and Watts-Tobin, R.J. (1961). 'General nature of the genetic code for proteins', *Nature, Lond.*, 192, 1227.

Goodman, H.M., Abelson, J., Landy, A., Brenner, S. and Smith, J.D. (1968). 'Amber suppression: a nucleotide change in the anticodon of a tyrosine transfer RNA', *Nature*, 217, 1019.

Hartman, P.E., Hartman, Z., and Serman, D. (1960). 'Complementation mapping by abortive transduction of histidine requiring Salmonella mutants', *J. Gen. Microbiol.*, 22, 354.

Hartman, P.E., Loper, J.C., and Serman, D. (1960). 'Fine structure mapping by complete transduction between histidine requiring Salmonella mutants', *J. Gen., Microbiol.*, 22, 323.

Stewart, J.W., Sherman, F., Jackson, M., Thomas, F.L.X. and Shipman, N. (1972). 'Demonstration of the UAA ochre codon in baker's yeast by amino-acid replacements in iso-1 cytochrome c. *J. Mol. Biol.*, 68, 83. (Also contains references to earlier papers dealing with the same classic system.)

Yanofsky, C., Carlton, B.C., Guest, J.R., Helinski, D.R. and Henning, U. (1964). 'On the collinearity of gene structure and protein structure', *Proc. Nat. Acad. Sci., U.S.A.*, 51, 266.

Yanofsky, C., Ito, J. and Horn, V. (1966). 'Amino acid replacements and the genetic code', *Cold Spring Harb. Symp. Quant. Biol.*, 31, 151. (A review of some of the data on *E. coli* tryptophan synthetase.)

Chapter 6

Ames, B.N. and Garry, B. (1959). 'Co-ordinate repression of the synthesis of four histidine biosynthetic enzymes by histidine', *Proc. Nat. Acad. Sci., U.S.A.*, 45, 1453.

Ames, B.N. and Hartman, P.E. (1963)—see references to Chapter 5.

Beckwith, J.R. and Zipser, D. (1970). *The Lactose Operon,* pp. 437. Cold Spring Harbor Laboratory, N.Y. (A comprehensive account of the best analysed operon.)

Gilbert, W. and Maxam, A. (1973). 'The nucleotide sequence of the *lac* operator'. *Proc. Nat. Acad. Sci., U.S.A.,* 70, 3581.

Hershey, A.D. (Editor) (1971). *The Bacteriophage Lambda,* pp. 792. Cold Spring Harbor Laboratory, N.Y.

Herskowitz, I. (1973). 'Control of gene expression in bacteriophage lambda', *Ann. Rev. Genet.,* 7, 289.

Jacob, F., and Monod, J. (1961). 'Genetic regulatory mechanisms in synthesis of proteins', *J. Mol. Biol.,* 3, 318. (A clear statement of the original operon model.)

Maizels, N. (1973). 'The nucleotide sequence of the lactose messenger ribonucleic acid transcribed from the UV5 promoter mutant of *Escherichia coli',* *Proc. Nat. Acad. Sci. U.S.A.,* 70, 3585.

Vogel, H.J. (Editor) (1971). *Metabolic Regulation* (Vol. 5 of *Metabolic Pathways,* third edition), pp. 576. Academic Press, N.Y. and London. (Contains authoritative reviews of all the systems mentioned in this chapter.)

Chapter 7

Margulis, L. (1970). *Origin of Eukaryotic Cells,* Yale University Press, New Haven, Connecticut. (Putting the case for the independent prokaryotic origin of organelles of eukaryotes.)

Meynell, G.G. (1972). *Bacterial Plasmids. Conjugation, Lysogeny and Transmissible Drug Resistance,* pp. 164 + xiii Macmillan, London.

Novick, R.P. (1969). 'Extrachromosomal inheritance in bacteria', *Bact. Rev.* 32, 210-235. (Probably still the best general concise review of plasmids.)

Ozeki, H., Howart, S., and Clowes, R.C. (1961). 'Colicin factors as fertility factors in bacteria', *Nature,* 190, 986.

Watanabe, T. (1963). 'Episome-mediated transfer of drug resistance in Entero-bacteriacea. VI. High-frequency resistance transfer system in *Escherichia coli',* *J. Bacteriol.,* 85, 788.

Chapter 8

Agarwhal, K.C., Khorana, H.G. and eleven others (1970). 'Total synthesis of the gene for an alanine transfer ribonucleic acid from yeast', *Nature,* 227, 27.

Allet, B., Jeppesen, P.G.N., Katagiri, K.J. and Delius, H. (1973). 'Mapping the DNA fragments produced by cleavage of λ DNA with endonuclease RI', *Nature,* 241, 120.

Ashby, Lord (Chairman). (1975). Report of Working Party on the Experimental Manipulation of Micro-organisms. H.M.S.O.

Berg, P. *et al.* (1974). Statement by a committee of the U.S. National Academy of Sciences on genetic engineering. *Nature,* 250, 175.

Cohen, S.N., Chang, A.C.Y., Boyer, H.W. and Helling, R.B. (1973). 'Construction of biologically functional bacterial plasmids *in vitro'*, *Proc. Nat. Acad. Sci. U.S.A.* 70, 3240.

Davis, R.W. and Parkinson, J.S. (1971). 'Deletion mutants of bacteriophage *lambda III*. Physical structure of *att*ϕ', *J. Mol. Biol.* 56, 403.

Hyman, R.W. and Summers, W.C. (1972). 'Isolation and physical mapping of T7 gene 1 messenger RNA', *J. Mol. Biol.* 71, 573.

Inman, R.B. and Schnös, M. (1970). 'Partial denaturation of thymine- and 5-bromo-uracil-containing λDNA in alkali', *J. Mol. Biol.* 49, 93.

Kacian, D.L., Spiegelman, S., Bank, H., Terada, M., Metafora, S., Dow, L. and Marks, P.A. (1972). '*In vitro* synthesis of DNA components of human genes for globins', *Nature New Biol.,* 235, 167.

Khorana, H.G. and thirteen others (1972). 'Total synthesis of the structural gene for an alanine transfer ribonucleic acid in yeast', *J. Mol. Biol.,* 72, 209.

Morrow, J.F., Cohen, S.N., Chang, A.C.Y., Boyer, H.W., Goodman, H.M. and Helling, R.B. (1974). 'Replication and transcription of eukaryotic DNA in *Escherichia coli*', *Proc. Nat. Acad. Sci. U.S.A.,* 71, 1743. (By construction of an *E. coli* plasmid incorporating toad rRNA genes.)

Parkinson, J.S. and Davis, R.W. (1971). 'A physical map of the left arm of the lambda chromosome', *J. Mol. Biol.* 56, 425.

Petes, T.D., Byers, B. and Fangman, W.L. (1973). 'Size and structure of yeast chromosomal DNA', *Proc. Nat. Acad. Sci. U.S.A.,* 70, 3072.

Price, P., Conover, J. and Hirschhorn, K. (1972). 'Chromosomal localization of human hemoglobin structural genes', *Nature,* 237, 340.

Shapiro, J., MacHattie, L., Eron, L., Ihler, G., Ippen, K. and Beckwith, J. (1969). 'Isolation of pure *lac* operon DNA', *Nature,* 224, 786.

Shih, T.Y. and Martin, M.A. (1973). 'A general method of gene isolation', *Proc. Nat. Acad. Sci. U.S.A.,* 70, 1697.

Varmus, H.E., Vogt, P.K. and Bishop, J.M. (1973). 'Integration of deoxyribonucleic acid specific for Rous sarcoma virus after infection of permissive and nonpermissive hosts', *Proc. Nat. Acad. Sci. U.S.A.,* 70, 3067.

Wu, M., Davidson, N. and Carbon, J. (1973). 'Physical mapping of the transfer RNA genes in $\lambda h80dglyTsu^{+}_{36}$', *J. Mol. Biol.,* 78, 23.

Yankowsky, S.A. and Spiegelman, S. (1962). 'The identification of the ribosomal RNA cistron by sequence complementarity.' *Proc. Nat. Acad. Sci. U.S.A.,* 48, 1069.

General references

(Dealing with the same material as this book, but more comprehensively and at a more advanced level.)

Hayes, W. (1968). *The Genetics of Bacteria and their Viruses.* Second edition, pp. 925. Blackwell Scientific Publ., Oxford.

Stent, F. (1963). *The Molecular Biology of the Bacterial Viruses*, pp. 473. W.H. Freeman and Co., San Francisco.

Stent, F. (1971). *Molecular Genetics, an Introductory Narrative*, pp. 650. W.H. Freeman and Co., San Francisco.

Watson, J.D. (1970). *The Molecular Biology of the Gene.* Second edition. pp. 662. W.A. Benjamin, New York.

Index